INORGANIC CHEMISTRY

Gary L. Miessler · Donald A. Tarr

—— Custom Edition for the University of Windsor ——

Taken from:

Inorganic Chemistry, Third Edition
by Gary L. Miessler and Donald A. Tarr

Cover image: courtesy of PhotoDisc/Getty Images

Taken from:

Inorganic Chemistry, Third Edition
by Gary L. Miessler and Donald A. Tarr
Copyright © 2004, 1999, 1991 by Pearson Education, Inc.
Published by Prentice-Hall, Inc.
Upper Saddle River, New Jersey 07458

This special edition published in cooperation with Pearson Custom Publishing.

Printed in Canada

10 9 8 7 6 5 4

ISBN 0-536-30614-1

2006180161

AG

Please visit our web site at *www.pearsoncustom.com*

PEARSON CUSTOM PUBLISHING
75 Arlington Street, Suite 300, Boston, MA 02116
A Pearson Education Company

Contents

CHAPTER

4

Symmetry and Group Theory

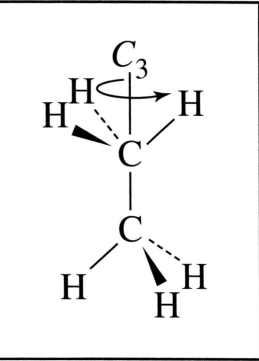

Symmetry is a phenomenon of the natural world, as well as the world of human invention (Figure 4-1). In nature, many types of flowers and plants, snowflakes, insects, certain fruits and vegetables, and a wide variety of microscopic plants and animals exhibit characteristic symmetry. Many engineering achievements have a degree of symmetry that contributes to their esthetic appeal. Examples include cloverleaf intersections, the pyramids of ancient Egypt, and the Eiffel Tower.

Symmetry concepts can be extremely useful in chemistry. By analyzing the symmetry of molecules, we can predict infrared spectra, describe the types of orbitals used in bonding, predict optical activity, interpret electronic spectra, and study a number of additional molecular properties. In this chapter, we first define symmetry very specifically in terms of five fundamental symmetry operations. We then describe how molecules can be classified on the basis of the types of symmetry they possess. We conclude with examples of how symmetry can be used to predict optical activity of molecules and to determine the number and types of infrared-active stretching vibrations.

In later chapters, symmetry will be a valuable tool in the construction of molecular orbitals (Chapters 5 and 10) and in the interpretation of electronic spectra of coordination compounds (Chapter 11) and vibrational spectra of organometallic compounds (Chapter 13).

A molecular model kit is a very useful study aid for this chapter, even for those who can visualize three-dimensional objects easily. We strongly encourage the use of such a kit.

4-1
SYMMETRY ELEMENTS AND OPERATIONS

All molecules can be described in terms of their symmetry, even if it is only to say they have none. Molecules or any other objects may contain **symmetry elements** such as mirror planes, axes of rotation, and inversion centers. The actual reflection, rotation, or inversion is called the **symmetry operation**. To contain a given symmetry element, a molecule must have exactly the same appearance after the operation as before. In other words, photographs of the molecule (if such photographs were possible!) taken from the same location before and after the symmetry operation would be indistinguishable. If a symmetry operation yields a molecule that can be distinguished from the original in

FIGURE 4-1 Symmetry in Nature, Art, and Architecture.

any way, then that operation is *not* a symmetry operation of the molecule. The examples in Figures 4-2 through 4-6 illustrate the possible types of molecular symmetry operations and elements.

The **identity operation (E)** causes no change in the molecule. It is included for mathematical completeness. An identity operation is characteristic of every molecule, even if it has no other symmetry.

The **rotation operation (C_n)** (also called **proper rotation**) is rotation through $360°/n$ about a rotation axis. We use counterclockwise rotation as a positive rotation. An example of a molecule having a threefold (C_3) axis is $CHCl_3$. The rotation axis is coincident with the C—H bond axis, and the rotation angle is $360°/3 = 120°$. Two C_3 operations may be performed consecutively to give a new rotation of $240°$. The resulting

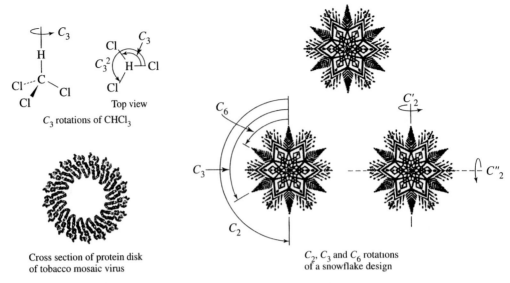

FIGURE 4-2 Rotations. The cross section of the tobacco mosaic virus is a cover diagram from *Nature*, 1976, *259*. © **1976**, Macmillan Journals Ltd. Reproduced with permission of Aaron Klug.

operation is designated $C_3{}^2$ and is also a symmetry operation of the molecule. Three successive C_3 operations are the same as the identity operation ($C_3{}^3 \equiv E$). The identity operation is included in all molecules. Many molecules and other objects have multiple rotation axes. Snowflakes are a case in point, with complex shapes that are nearly always hexagonal and nearly planar. The line through the center of the flake perpendicular to the plane of the flake contains a twofold (C_2) axis, a threefold (C_3) axis, and a sixfold (C_6) axis. Rotations by 240° ($C_3{}^2$) and 300° ($C_6{}^5$) are also symmetry operations of the snowflake.

Rotation Angle	Symmetry Operation
60°	C_6
120°	C_3 ($\equiv C_6{}^2$)
180°	C_2 ($\equiv C_6{}^3$)
240°	$C_3{}^2$ ($\equiv C_6{}^4$)
300°	$C_6{}^5$
360°	E ($\equiv C_6{}^6$)

There are also two sets of three C_2 axes in the plane of the snowflake, one set through opposite points and one through the cut-in regions between the points. One of each of these axes is shown in Figure 4-2. In molecules with more than one rotation axis, the C_n axis having the largest value of n is the **highest order rotation axis** or **principal axis**. The highest order rotation axis for a snowflake is the C_6 axis. (In assigning Cartesian coordinates, the highest order C_n axis is usually chosen as the z axis.) When necessary, the C_2 axes perpendicular to the principal axis are designated with primes; a single prime ($C_2{}'$) indicates that the axis passes through several atoms of the molecule, whereas a double prime ($C_2{}''$) indicates that it passes between the outer atoms.

Finding rotation axes for some three-dimensional figures is more difficult, but the same in principle. Remember that nature is not always simple when it comes to symmetry—the protein disk of the tobacco mosaic virus has a 17-fold rotation axis!

In the **reflection operation** (σ) the molecule contains a mirror plane. If details such as hair style and location of internal organs are ignored, the human body has a left-right mirror plane, as in Figure 4-3. Many molecules have mirror planes, although they may not be immediately obvious. The reflection operation exchanges left and right, as if each point had moved perpendicularly through the plane to a position exactly as far from the plane as when it started. Linear objects such as a round wood pencil or molecules

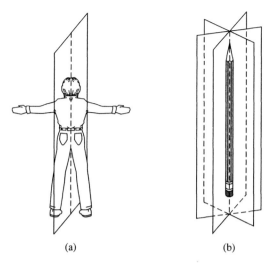

FIGURE 4-3 Reflections.

     (a)   (b)

such as acetylene or carbon dioxide have an infinite number of mirror planes that include the center line of the object.

When the plane is perpendicular to the principal axis of rotation, it is called σ_h (horizontal). Other planes, which contain the principal axis of rotation, are labeled σ_v or σ_d.

Inversion (i) is a more complex operation. Each point moves through the center of the molecule to a position opposite the original position and as far from the central point as when it started.[1] An example of a molecule having a center of inversion is ethane in the staggered conformation, for which the inversion operation is shown in Figure 4-4.

Many molecules that seem at first glance to have an inversion center do not; for example, methane and other tetrahedral molecules lack inversion symmetry. To see this, hold a methane model with two hydrogen atoms in the vertical plane on the right and two hydrogen atoms in the horizontal plane on the left, as in Figure 4-4. Inversion results in two hydrogen atoms in the horizontal plane on the right and two hydrogen atoms in the vertical plane on the left. Inversion is therefore *not* a symmetry operation of methane, because the orientation of the molecule following the i operation differs from the original orientation.

Squares, rectangles, parallelograms, rectangular solids, octahedra, and snowflakes have inversion centers; tetrahedra, triangles, and pentagons do not (Figure 4-5).

A **rotation-reflection operation (S_n)** (sometimes called **improper rotation**) requires rotation of $360°/n$, followed by reflection through a plane perpendicular to the axis of rotation. In methane, for example, a line through the carbon and bisecting the

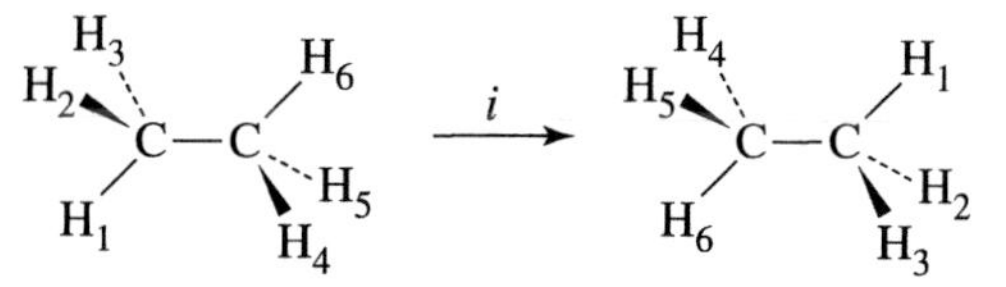

$$H_2 \overset{H_3}{\underset{H_1}{\diagdown}} C{-}C \overset{H_6}{\underset{H_4}{\diagup}} H_5 \quad \xrightarrow{\ i\ } \quad H_5 \overset{H_4}{\underset{H_6}{\diagdown}} C{-}C \overset{H_1}{\underset{H_3}{\diagup}} H_2$$

Center of inversion

$$H_1 \overset{H_3}{\underset{H_2}{\diagdown}} C \overset{}{\underset{H_4}{\diagdown}} \quad \xrightarrow{\ i\ } \quad \overset{H_4}{\underset{H_3}{}} C \overset{H_2}{\underset{H_1}{\diagup}}$$

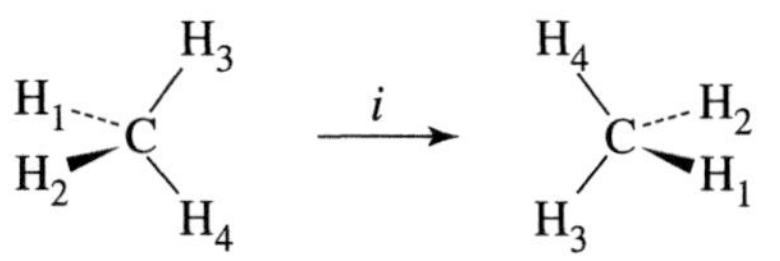

No center of inversion

FIGURE 4-4 Inversion.

[1]This operation must be distinguished from the inversion of a tetrahedral carbon in a bimolecular reaction, which is more like that of an umbrella in a high wind.

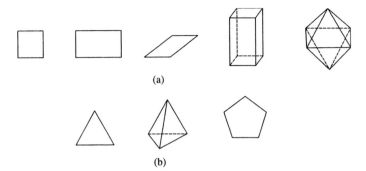

FIGURE 4-5 Figures (a) With and (b) Without Inversion Centers.

angle between two hydrogen atoms on each side is an S_4 axis. There are three such lines, for a total of three S_4 axes. The operation requires a 90° rotation of the molecule followed by reflection through the plane perpendicular to the axis of rotation. Two S_n operations in succession generate a $C_{n/2}$ operation. In methane, two S_4 operations generate a C_2. These operations are shown in Figure 4-6, along with a table of C and S equivalences for methane.

Molecules sometimes have an S_n axis that is coincident with a C_n axis. For example, in addition to the rotation axes described previously, snowflakes have S_2 ($\equiv i$), S_3, and S_6 axes coincident with the C_6 axis. Molecules may also have S_{2n} axes coincident with C_n; methane is an example, with S_4 axes coincident with C_2 axes, as shown in Figure 4-6.

Note that an S_2 operation is the same as inversion; an S_1 operation is the same as a reflection plane. The i and σ notations are preferred in these cases. Symmetry elements and operations are summarized in Table 4-1.

Rotation Angle	Symmetry Operation
90°	S_4
180°	C_2 $(\equiv S_4{}^2)$
270°	$S_4{}^3$
360°	E $(\equiv S_4{}^4)$

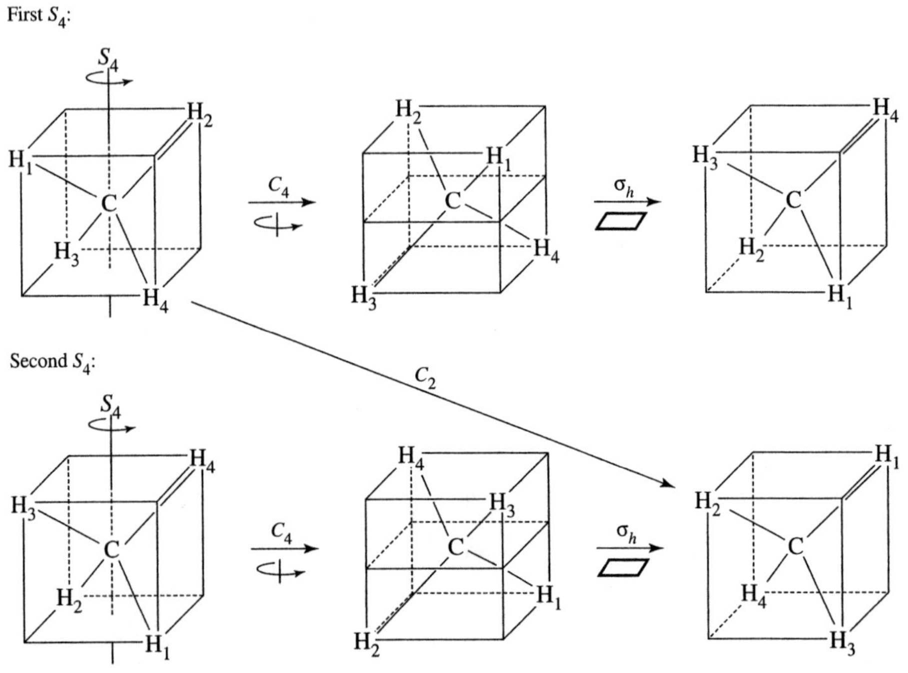

First S_4:

Second S_4:

FIGURE 4-6 Improper Rotation or Rotation-Reflection.

TABLE 4-1
Summary Table of Symmetry Elements and Operations

Symmetry Operation	Symmetry Element	Operation	Examples
Identity, E	None	All atoms unshifted	CHFClBr
Rotation, C_2	Rotation axis	Rotation by $360°/n$	p-dichlorobenzene
C_3			NH_3
C_4			$[PtCl_4]^{2-}$
C_5			Cyclopentadienyl group
C_6			Benzene
Reflection, σ	Mirror plane	Reflection through a mirror plane	H_2O
Inversion, i	Inversion center (point)	Inversion through the center	Ferrocene (staggered)
Rotation-reflection, S_4	Rotation-reflection axis (improper axis)	Rotation by $360°/n$, followed by reflection in the plane perpendicular to the rotation axis	CH_4
S_6			Ethane (staggered)
S_{10}			Ferrocene (staggered)

EXAMPLES

Find all the symmetry elements in the following molecules; consider only the atoms when assigning symmetry. Lone pairs influence shapes, but molecular symmetry is based on the geometry of the atoms.

H_2O

H_2O has two planes of symmetry, one in the plane of the molecule and one perpendicular to the molecular plane, as shown in Table 4-1. It also has a C_2 axis collinear with the intersection of the mirror planes. H_2O has no inversion center.

p-Dichlorobenzene   This molecule has three mirror planes: the molecular plane; a plane perpendicular to the molecule, passing through both chlorines; and a plane perpendicular to the first two, bisecting the molecule between the chlorines. It also has three C_2 axes, one perpendicular to the molecular plane (see Table 4-1) and two within the plane: one passing through both chlorines and one perpendicular to the axis passing through the chlorines. Finally, p-dichlorobenzene has an inversion center.

Ethane (staggered conformation)   Ethane has three mirror planes, each containing the C—C bond axis and passing through two hydrogens on opposite ends of the molecule. It has a C_3 axis collinear with the carbon-carbon bond and three C_2 axes bisecting the angles between the mirror planes. (Use of a model is especially helpful for viewing the C_2 axes). Ethane also has a center of inversion and an S_6 axis collinear with the C_3 axis (see Table 4-1).

EXERCISE 4-1
Using diagrams as necessary, show that $S_2 \equiv i$ and $S_1 \equiv \sigma$.

EXERCISE 4-2

Find all the symmetry elements in the following molecules:

NH_3 Cyclohexane (boat conformation)   Cyclohexane (chair conformation) XeF_2

4-2
POINT GROUPS

Each molecule has a set of symmetry operations that describes the molecule's overall symmetry. This set of symmetry operations is called the **point group** of the molecule. **Group theory**, the mathematical treatment of the properties of groups, can be used to determine the molecular orbitals, vibrations, and other properties of the molecule. With only a few exceptions, the rules for assigning a molecule to a point group are simple and straightforward. We need only to follow these steps in sequence until a final classification of the molecule is made. A diagram of these steps is shown in Figure 4-7.

1.   Determine whether the molecule belongs to one of the cases of very low symmetry (C_1, C_s, C_i) or high symmetry $(T_d, O_h, C_{\infty v}, D_{\infty h},$ or $I_h)$ described in Tables 4-2 and 4-3.

2.   For all remaining molecules, find the rotation axis with the highest n, the highest order C_n axis for the molecule.

3.   Does the molecule have any C_2 axes perpendicular to the C_n axis? If it does, there will be n of such C_2 axes, and the molecule is in the D set of groups. If not, it is in the C or S set.

4.   Does the molecule have a mirror plane (σ_h) perpendicular to the C_n axis? If so, it is classified as C_{nh} or D_{nh}. If not, continue with Step 5.

5.   Does the molecule have any mirror planes that contain the C_n axis $(\sigma_v$ or $\sigma_d)$? If so, it is classified as C_{nv} or D_{nd}. If not, but it is in the D set, it is classified as D_n. If the molecule is in the C or S set, continue with Step 6.

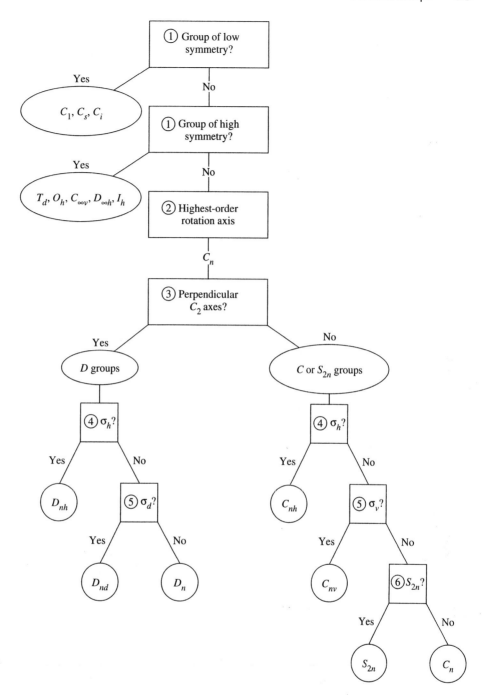

FIGURE 4-7 Diagram of the Point Group Assignment Method.

6. Is there an S_{2n} axis collinear with the C_n axis? If so, it is classified as S_{2n}. If not, the molecule is classified as C_n.

Each step is illustrated in the following text by assigning the molecules in Figure 4-8 to their point groups. The low- and high-symmetry cases are treated differently because of their special nature. Molecules that are not in one of these low- or high-symmetry point groups can be assigned to a point group by following Steps 2 through 6.

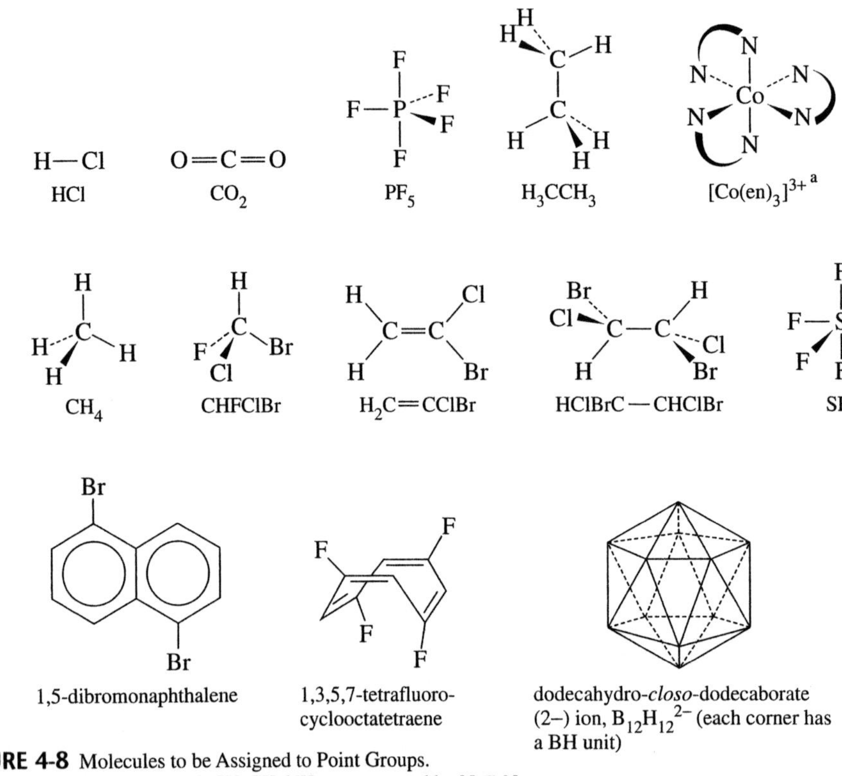

FIGURE 4-8 Molecules to be Assigned to Point Groups.
[a]en = ethylenediamine = $NH_2CH_2CH_2NH_2$, represented by N⌒N.

4-2-1 GROUPS OF LOW AND HIGH SYMMETRY

1. Determine whether the molecule belongs to one of the special cases of low or high symmetry.

First, inspection of the molecule will determine if it fits one of the low-symmetry cases. These groups have few or no symmetry operations and are described in Table 4-2.

TABLE 4-2
Groups of Low Symmetry

Group	Symmetry	Examples	
C_1	No symmetry other than the identity operation	CHFClBr	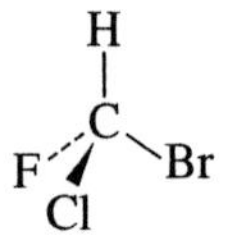
C_s	Only one mirror plane	H_2C=CClBr	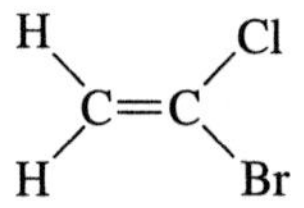
C_i	Only an inversion center; few molecular examples	HClBrC—CHClBr (staggered conformation)	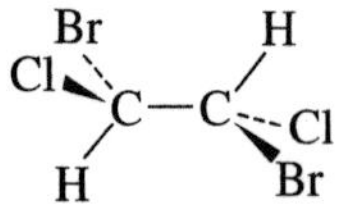

Low symmetry

CHFClBr has no symmetry other than the identity operation and has C_1 symmetry, H_2C=CClBr has only one mirror plane and C_s symmetry, and HClBrC—CHClBr in the conformation shown has only a center of inversion and C_i symmetry.

High symmetry

Molecules with many symmetry operations may fit one of the high-symmetry cases of linear, tetrahedral, octahedral, or icosahedral symmetry with the characteristics described in Table 4-3. Molecules with very high symmetry are of two types, linear and polyhedral. Linear molecules having a center of inversion have $D_{\infty h}$ symmetry; those lacking an inversion center have $C_{\infty v}$ symmetry. The highly symmetric point groups T_d, O_h, and I_h are described in Table 4-3. It is helpful to note the C_n axes of these molecules. Molecules with T_d symmetry have only C_3 and C_2 axes; those with O_h symmetry have C_4 axes in addition to C_3 and C_2; and I_h molecules have C_5, C_3, and C_2 axes.

TABLE 4-3
Groups of High Symmetry

Group	Description	Examples
$C_{\infty v}$	These molecules are linear, with an infinite number of rotations and an infinite number of reflection planes containing the rotation axis. They do not have a center of inversion.	C_∞ H—Cl
$D_{\infty h}$	These molecules are linear, with an infinite number of rotations and an infinite number of reflection planes containing the rotation axis. They also have perpendicular C_2 axes, a perpendicular reflection plane, and an inversion center.	C_∞ O=C=O C_2
T_d	Most (but not all) molecules in this point group have the familiar tetrahedral geometry. They have four C_3 axes, three C_2 axes, three S_4 axes, and six σ_d planes. They have no C_4 axes.	H C—H H H
O_h	These molecules include those of octahedral structure, although some other geometrical forms, such as the cube, share the same set of symmetry operations. Among their 48 symmetry operations are four C_3 rotations, three C_4 rotations, and an inversion.	F F F—S—F F F
I_h	Icosahedral structures are best recognized by their six C_5 axes (as well as many other symmetry operations—120 total).	$B_{12}H_{12}{}^{2-}$ with BH at each vertex of an icosahedron

In addition, there are four other groups, T, T_h, O, and I, which are rarely seen in nature. These groups are discussed at the end of this section.

HCl has $C_{\infty v}$ symmetry, CO_2 has $D_{\infty h}$ symmetry, CH_4 has tetrahedral (T_d) symmetry, SF_6 has octahedral (O_h) symmetry, and $B_{12}H_{12}{}^{2-}$ has icosahedral (I_h) symmetry

There are now seven molecules left to be assigned to point groups out of the original 15.

4-2-2 OTHER GROUPS

> 2. Find the rotation axis with the highest n, the highest order C_n axis for the molecule. This is the principal axis of the molecule.

The rotation axes for the examples are shown in Figure 4-9. If they are all equivalent, any one can be chosen as the principal axis.

PF_5
H_3CCH_3
1,3,5,7-tetrafluoro-cyclooctatetraene
C_3 perpendicular to the plane of the page
$[Co(en)_3]^{3+}$
NH_3
C_2 perpendicular to the plane of the molecule
1,5-dibromonaphthalene
H_2O_2

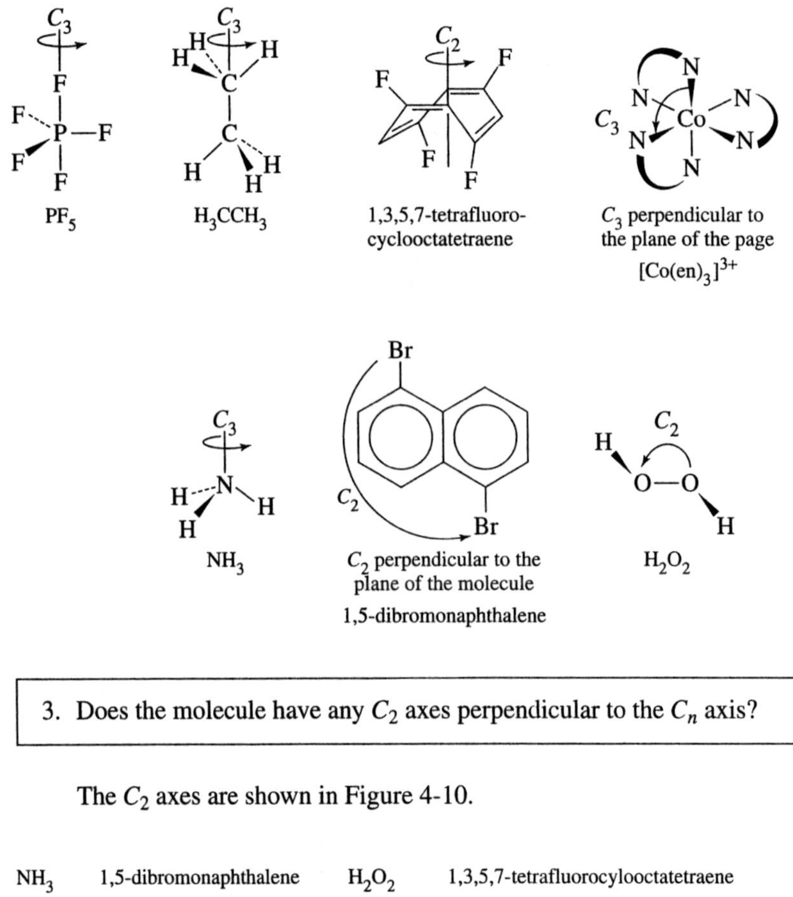

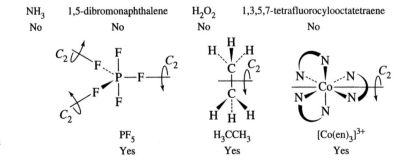

FIGURE 4-9 Rotation Axes.

> 3. Does the molecule have any C_2 axes perpendicular to the C_n axis?

The C_2 axes are shown in Figure 4-10.

NH_3 No
1,5-dibromonaphthalene No
H_2O_2 No
1,3,5,7-tetrafluorocylooctatetraene No
PF_5 Yes
H_3CCH_3 Yes
$[Co(en)_3]^{3+}$ Yes

FIGURE 4-10 Perpendicular C_2 Axes.

Yes *D Groups*

PF_5, H_3CCH_3, $[Co(en)_3]^{3+}$

Molecules with C_2 axes perpendicular to the principal axis are in one of the groups designated by the letter D; there are n C_2 axes.

No *C or S Groups*

NH_3, 1,5-dibromonaphthalene, H_2O_2, 1,3,5,7-tetrafluorocyclooctatetraene

Molecules with no perpendicular C_2 axes are in one of the groups designated by the letters C or S.

No final assignments of point groups have been made, but the molecules have now been divided into two major categories, the D set and the C or S set.

---

4. Does the molecule have a mirror plane (σ_h horizontal plane) perpendicular to the C_n axis?

The horizontal mirror planes are shown in Figure 4-11.

D Groups

H_3CCH_3	$[Co(en)_3]^{3+}$
No	No

C and *S* Groups

NH_3	H_2O_2	1,3,5,7-tetrafluoro-cyclooctatetraene
No	No	No

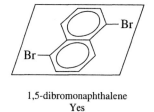

PF$_5$
Yes
D$_{3h}$

1,5-dibromonaphthalene
Yes
C$_{2h}$

FIGURE 4-11 Horizontal Mirror Planes.

D Groups

Yes $\boxed{D_{nh}}$

PF$_5$ is D_{3h}

C and S Groups

Yes $\boxed{C_{nh}}$

1,5-dibromonaphthalene is C_{2h}

These molecules are now assigned to point groups and need not be considered further. Both have horizontal mirror planes.

No D_n *or* D_{nd}

H_3CCH_3, $[Co(en)_3]^{3+}$

No C_n, C_{nv}, *or* S_{2n}

NH_3, H_2O_2,
1,3,5,7-tetrafluorocyclooctatetraene

None of these have horizontal mirror planes; they must be carried further in the process.

---

5. Does the molecule have any mirror planes that contain the C_n axis?

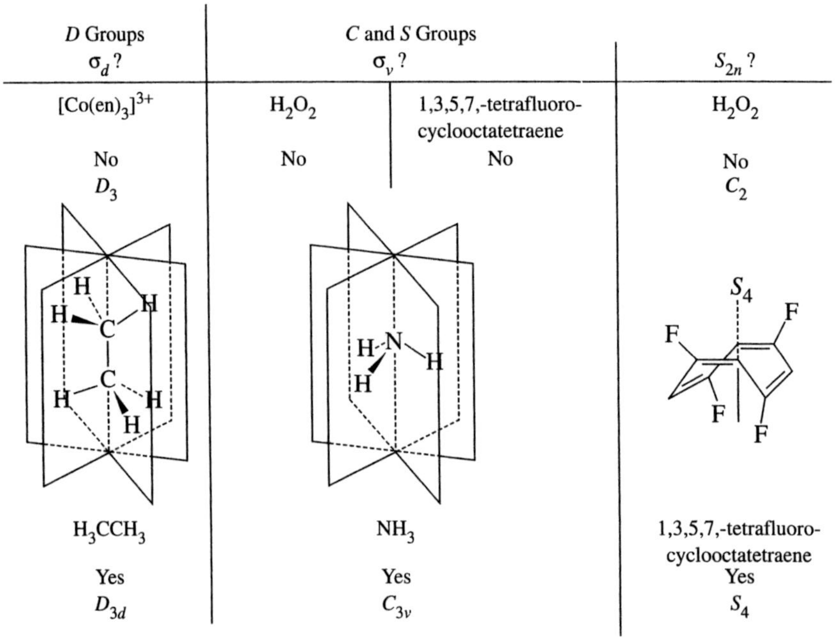

D Groups σ_d?	C and S Groups σ_v?		S_{2n}?
$[Co(en)_3]^{3+}$	H_2O_2	1,3,5,7,-tetrafluoro-cyclooctatetraene	H_2O_2
No D_3	No	No	No C_2

H_3CCH_3 Yes D_{3d}	NH_3 Yes C_{3v}	1,3,5,7,-tetrafluoro-cyclooctatetraene Yes S_4

FIGURE 4-12 Vertical or Dihedral Mirror Planes or S_{2n} Axes.

These mirror planes are shown in Figure 4-12.

D Groups

Yes $\boxed{D_{nd}}$

H_3CCH_3 (staggered) is D_{3d}

C and S Groups

Yes $\boxed{C_{nv}}$

NH_3 is C_{3v}

These molecules have mirror planes containing the major C_n axis, but no horizontal mirror planes, and are assigned to the corresponding point groups. There will be n of these planes.

No $\boxed{D_n}$

$[Co(en)_3]^{3+}$ is D_3

No C_n or S_{2n}

H_2O_2, 1,3,5,7-tetrafluorocyclooctate-traene

These molecules are in the simpler rotation groups D_n, C_n, and S_{2n} because they do not have any mirror planes. D_n and C_n point groups have *only* C_n axes. S_{2n} point groups have C_n and S_{2n} axes and may have an inversion center.

---

6. Is there an S_{2n} axis collinear with the C_n axis?

D Groups

Any molecules in this category that have S_{2n} axes have already been assigned to groups. There are no additional groups to be considered here.

C and S Groups

Yes $\boxed{S_{2n}}$

1,3,5,7-tetrafluorocyclooctatetraene is S_4

No $\boxed{C_n}$

H_2O_2 is C_2

We have only one example in our list that falls into the S_{2n} groups, as seen in Figure 4-12.

A branching diagram that summarizes this method of assigning point groups was given in Figure 4-7 and more examples are given in Table 4-4.

TABLE 4-4
Further Examples of *C* and *D* Point Groups

General Label	Point Group and Example	
C_{nh}	C_{2h}	difluorodiazene
	C_{3h}	$B(OH)_3$, planar
C_{nv}	C_{2v}	H_2O
	C_{3v}	PCl_3
	C_{4v}	BrF_5 (square pyramid)
	$C_{\infty v}$	HF, CO, HCN
C_n	C_2	N_2H_4, which has a *gauche* conformation
	C_3	$P(C_6H_5)_3$, which is like a three-bladed propeller distorted out of the planar shape by a lone pair on the P
D_{nh}	D_{3h}	BF_3
	D_{4h}	$PtCl_4{}^{2-}$
	D_{5h}	$Os(C_5H_5)_2$ (eclipsed)

Continued

TABLE 4-4—cont'd
Further Examples of C and D Point Groups

General Label	Point Group and Example	
	D_{6h} benzene	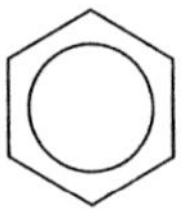
	$D_{\infty h}$ F_2, N_2,	F—F N≡N
	acetylene (C_2H_2)	H—C≡C—H
D_{nd}	D_{2d} $H_2C{=}C{=}CH_2$, allene	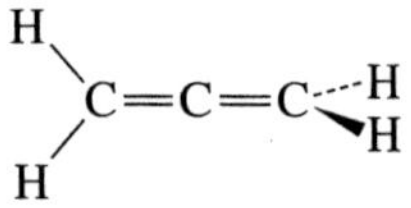
	D_{4d} Ni(cyclobutadiene)$_2$ (staggered)	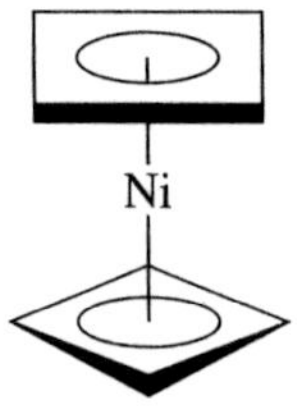
	D_{5d} Fe(C_5H_5)$_2$ (staggered)	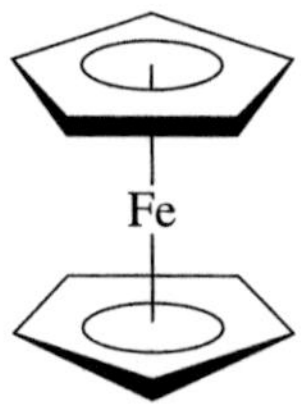
D_n	D_3 [Ru(NH$_2$CH$_2$CH$_2$NH$_2$)$_3$]$^{2+}$ (treating the NH$_2$CH$_2$CH$_2$NH$_2$ group as a planar ring)	

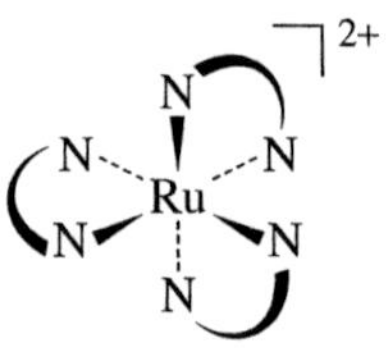

EXAMPLES

Determine the point groups of the following molecules and ions from Figures 3-13 and 3-16:

XeF$_4$ 1. XeF$_4$ is not in the groups of low or high symmetry.
2. Its highest order rotation axis is C_4.
3. It has four C_2 axes perpendicular to the C_4 axis and is therefore in the D set of groups.
4. It has a horizontal plane perpendicular to the C_4 axis. Therefore its point group is D_{4h}.

SF$_4$ 1. It is not in the groups of high or low symmetry.
2. Its highest order (and only) rotation axis is a C_2 axis passing through the lone pair.
3. The ion has no other C_2 axes and is therefore in the C or S set.
4. It has no mirror plane perpendicular to the C_2.
5. It has two mirror planes containing the C_2 axis. Therefore, the point group is C_{2v}.

IOF$_3$ 1. The molecule has no symmetry (other than E). Its point group is C_1.

EXERCISE 4-3

Use the procedure described above to verify the point groups of the molecules in Table 4-4.

C versus D point group classifications

All molecules having these classifications must have a C_n axis. If more than one C_n axis is found, the highest order axis (largest value of n) is used as the reference axis. In general, it is useful to orient this axis vertically.

	D Classifications	C Classifications
General Case: Look for C_2 axes perpendicular to the highest order C_n axis.	nC_2 axes $\perp C_n$ axis	No C_2 axes $\perp C_n$ axis
Subcategories:		
If a horizontal plane of symmetry exists:	D_{nh}	C_{nh}
If n vertical planes exist:	D_{nd}	C_{nv}
If no planes of symmetry exist:	D_n	C_n

Notes:

1. Vertical planes contain the highest order C_n axis. In the D_{nd} case, the planes are designated *dihedral* because they are between the C_2 axes—thus, the subscript d.
2. Simply having a C_n axis does not guarantee that a molecule will be in a D or C category; don't forget that the high-symmetry T_d, O_h, and I_h point groups and related groups have a large number of C_n axes.
3. When in doubt, you can always check the character tables (Appendix C) for a complete list of symmetry elements for any point group.

Groups related to I_h, O_h, and T_d groups

The high-symmetry point groups I_h, O_h, and T_d are well known in chemistry and are represented by such classic molecules as C_{60}, SF_6, and CH_4. For each of these point groups, there is also a purely rotational subgroup (I, O, and T, respectively) in which the only symmetry operations other than the identity operation are proper axes of rotation. The symmetry operations for these point groups are in Table 4-5.

We are not yet finished with high-symmetry point groups. One more group, T_h, remains. The T_h point group is derived by adding a center of inversion to the T point group; adding i generates the additional symmetry operations S_6, $S_6^{\,5}$, and σ_h.

TABLE 4-5
Symmetry Operations for High-Symmetry Point Groups and Their Rotational Subgroups

Point Group	Symmetry Operations									
I_h	E	$12C_5$	$12C_5^{\,2}$	$20C_3$	$15C_2$		i	$12S_{10}$	$12S_{10}^{\,3}$	$20S_6$	15σ
I	E	$12C_5$	$12C_5^{\,2}$	$20C_3$	$15C_2$						
O_h	E	$8C_3$	$6C_2$	$6C_4$	$3C_2\;(\equiv C_4^{\,2})$		i	$6S_4$	$8S_6$	$3\sigma_h$	$6\sigma_d$
O	E	$8C_3$	$6C_2$	$6C_4$	$3C_2\;(\equiv C_4^{\,2})$						
T_d	E	$8C_3$	$3C_2$	$6S_4$	$6\sigma_d$						
T	E	$4C_3\;4C_3^{\,2}$	$3C_2$								
T_h	E	$4C_3 4C_3^{\,2}$	$3C_2$				i	$4S_6$	$4S_6^{\,5}$	$3\sigma_h$	

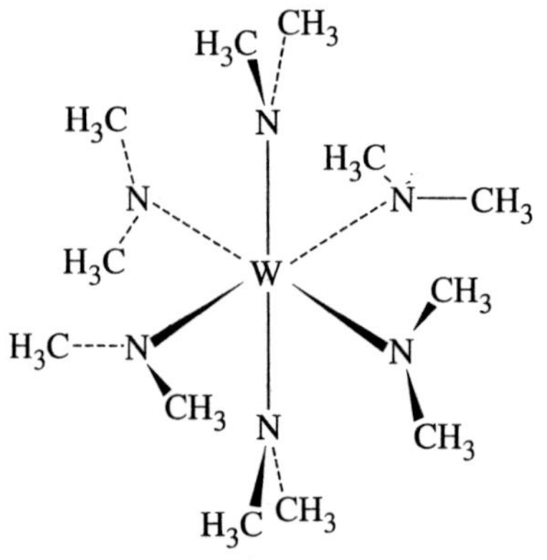

FIGURE 4-13 $W[N(CH_3)_2]_6$, a Molecule with T_h Symmetry.

T_h symmetry is rare but is known for a few molecules. The compound shown in Figure 4-13 is an example. I, O, and T symmetry are rarely if ever encountered in chemistry.

That's all there is to it! It takes a fair amount of practice, preferably using molecular models, to learn the point groups well, but once you know them, they can be extremely useful. Several practical applications of point groups appear later in this chapter, and additional applications are included in later chapters.

4-3 PROPERTIES AND REPRESENTATIONS OF GROUPS

All mathematical groups (of which point groups are special types) must have certain properties. These properties are listed and illustrated in Table 4-6, using the symmetry operations of NH_3 in Figure 4-14 as an example.

4-3-1 MATRICES

Important information about the symmetry aspects of point groups is summarized in character tables, described later in this chapter. To understand the construction and use of character tables, we first need to consider the properties of matrices, which are the basis for the tables.[2]

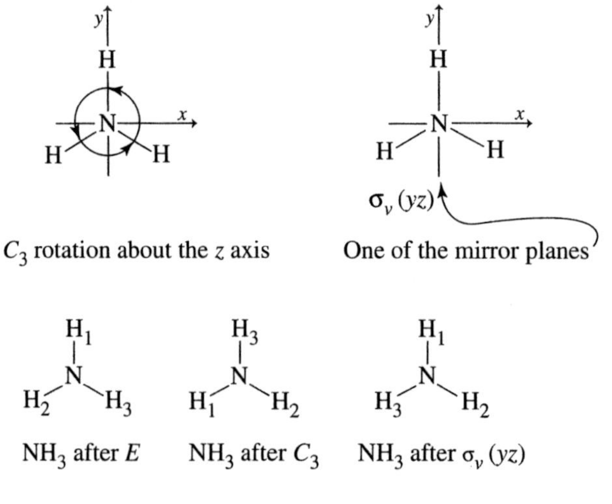

C_3 rotation about the z axis   One of the mirror planes

FIGURE 4-14 Symmetry Operations for Ammonia. (Top view) NH_3 is of point group C_{3v}, with the symmetry operations E, C_3, $C_3{}^2$, σ_v, $\sigma_v{}'$, $\sigma_v{}''$, usually written as E, $2C_3$, and $3\sigma_v$ (note that $C_3{}^3 \equiv E$).

NH_3 after E NH_3 after C_3 NH_3 after σ_v (yz)

[2]More details on matrices and their manipulation are available in Appendix 1 of F. A. Cotton, *Chemical Applications of Group Theory*, 3rd ed., John Wiley & Sons, New York, 1990, and in linear algebra and finite mathematics textbooks.

TABLE 4-6
Properties of a Group

Property of Group	Examples from Point Group C_{3v}

1. Each group must contain an **identity** operation that commutes (in other words, $EA = AE$) with all other members of the group and leaves them unchanged ($EA = AE = A$).

C_{3v} molecules (and *all* molecules) contain the identity operation E.

2. Each operation must have an **inverse** that, when combined with the operation, yields the identity operation (sometimes a symmetry operation may be its own inverse). *Note:* By convention, we perform combined symmetry operations *from right to left* as written.

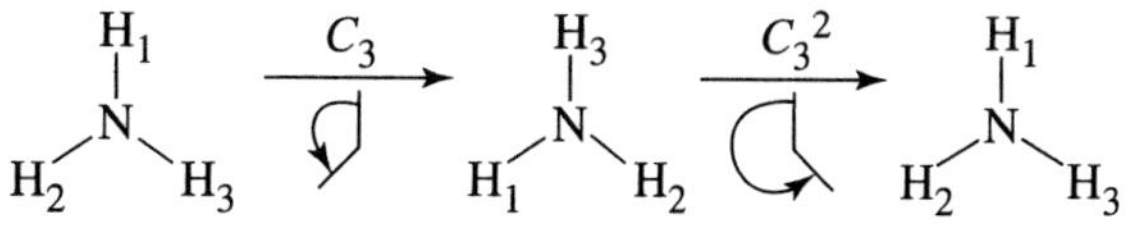

$C_3^2 C_3 = E$ (C_3 and C_3^2 are inverses of each other)

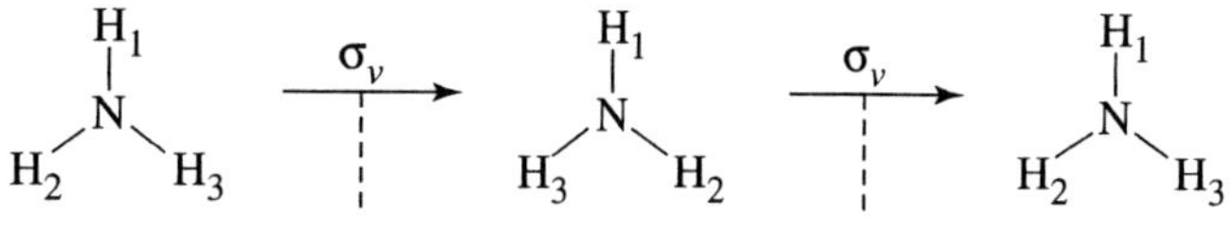

$\sigma_v \sigma_v = E$ (mirror planes are shown as dashed lines; σ_v is its own inverse)

3. The product of any two group operations must also be a member of the group. This includes the product of any operation with itself.

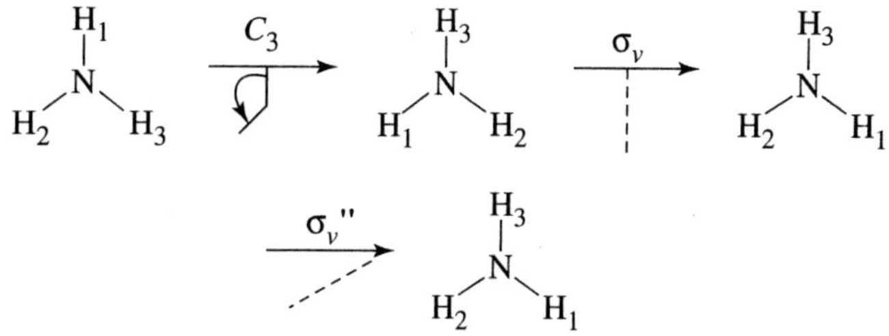

$\sigma_v C_3$ has the same overall effect as σ_v''; therefore, we write $\sigma_v C_3 = \sigma_v''$.

It can be shown that the products of any two operations in C_{3v} are also members of C_{3v}.

4. The associative property of combination must hold. In other words, $A(BC) = (AB)C$.

$C_3(\sigma_v \sigma_v') = (C_3 \sigma_v)\sigma_v'$

By **matrix** we mean an ordered array of numbers, such as

$$\begin{bmatrix} 3 & 2 \\ 7 & 1 \end{bmatrix} \quad \text{or} \quad [2 \ \ 0 \ \ 1 \ \ 3 \ \ 5]$$

To multiply matrices, it is first required that the number of vertical columns of the first matrix be equal to the number of horizontal rows of the second matrix. To find the product, sum, term by term, the products of each *row* of the first matrix by each *column* of the second (each term in a row must be multiplied by its corresponding term in the appropriate column of the second matrix). Place the resulting sum in the product matrix with the row determined by the row of the first matrix and the column determined by the column of the second matrix:

$$C_{ij} = \Sigma A_{ik} \times B_{kj}$$

Here

C_{ij} = product matrix, with i rows and j columns

A_{ik} = initial matrix, with i rows and k columns

B_{kj} = initial matrix, with k rows and j columns

EXAMPLES

$$i\begin{bmatrix} 1 & 5 \\ 2 & 6 \end{bmatrix} \times \begin{bmatrix} 7 & 3 \\ 4 & 8 \end{bmatrix}k = \begin{bmatrix} (1)(7) + (5)(4) & (1)(3) + (5)(8) \\ (2)(7) + (6)(4) & (2)(3) + (6)(8) \end{bmatrix}i = \begin{bmatrix} 27 & 43 \\ 38 & 54 \end{bmatrix}i$$

This example has 2 rows and 2 columns in each initial matrix, so it has 2 rows and 2 columns in the product matrix; $i = j = k = 2$.

$$i\begin{bmatrix} 1 & 2 & 3 \end{bmatrix}\begin{bmatrix} 1 & 0 & 0 \\ 0 & -1 & 0 \\ 0 & 0 & 1 \end{bmatrix}k =$$

$$[(1)(1) + (2)(0) + (3)(0) \quad (1)(0) + (2)(-1) + (3)(0) \quad (1)(0) + (2)(0) + (3)(1)]i = [1 \quad -2 \quad 3]i$$

Here, $i = 1$, $j = 3$, and $k = 3$, so the product matrix has 1 row (i) and 3 columns (j).

$$i\begin{bmatrix} 1 & 0 & 0 \\ 0 & -1 & 0 \\ 0 & 0 & 1 \end{bmatrix}\begin{bmatrix} 1 \\ 2 \\ 3 \end{bmatrix}k = \begin{bmatrix} (1)(1) + & (0)(2) + (0)(3) \\ (0)(1) + (-1)(2) + (0)(3) \\ (0)(1) + & (0)(2) + (1)(3) \end{bmatrix}i = \begin{bmatrix} 1 \\ -2 \\ 3 \end{bmatrix}i$$

Here $i = 3$, $j = 1$, and $k = 3$, so the product matrix has 3 rows (i) and 1 column (j).

EXERCISE 4-4

Do the following multiplications:

a. $\begin{bmatrix} 5 & 1 & 3 \\ 4 & 2 & 2 \\ 1 & 2 & 3 \end{bmatrix} \times \begin{bmatrix} 2 & 1 & 1 \\ 1 & 2 & 3 \\ 5 & 4 & 3 \end{bmatrix}$

b. $\begin{bmatrix} 1 & -1 & -2 \\ 0 & 1 & -1 \\ 1 & 0 & 0 \end{bmatrix} \times \begin{bmatrix} 2 \\ 1 \\ 3 \end{bmatrix}$

c. $\begin{bmatrix} 1 & 2 & 3 \end{bmatrix} \times \begin{bmatrix} 1 & -1 & -2 \\ 2 & 1 & -1 \\ 3 & 2 & 1 \end{bmatrix}$

4-3-2 REPRESENTATIONS OF POINT GROUPS

Symmetry operations: Matrix representations

Consider the effects of the symmetry operations of the C_{2v} point group on the set of x, y, and z coordinates. [The set of p orbitals (p_x, p_y, p_z) behaves the same way, so this is a useful exercise.] The water molecule is an example of a molecule having C_{2v} symmetry. It has a C_2 axis through the oxygen and in the plane of the molecule, no perpendicular C_2 axes, and no horizontal mirror plane, but it does have two vertical mirror planes,

Coordinate system	After C_2	After $\sigma_v(xz)$	After $\sigma_v{}'(yz)$

FIGURE 4-15 Symmetry Operations of the Water Molecule.

as shown in Table 4-1 and Figure 4-15. The z axis is usually chosen as the axis of highest rotational symmetry; for H_2O, this is the *only* rotational axis. The other axes are arbitrary. We will use the xz plane as the plane of the molecule.[3] This set of axes is chosen to obey the right-hand rule (the thumb and first two fingers of the right hand, held perpendicular to each other, are labeled x, y, and z, respectively).

Each symmetry operation may be expressed as a **transformation matrix** as follows:

$$[\text{New coordinates}] = [\text{transformation matrix}][\text{old coordinates}]$$

As examples, consider how transformation matrices can be used to represent the symmetry operations of the C_{2v} point group:

C_2: Rotate a point having coordinates (x, y, z) about the $C_2(z)$ axis. The new coordinates are given by

$$\begin{aligned} x' = \text{new } x &= -x \\ y' = \text{new } y &= -y \\ z' = \text{new } z &= z \end{aligned} \qquad \begin{bmatrix} -1 & 0 & 0 \\ 0 & -1 & 0 \\ 0 & 0 & 1 \end{bmatrix} \quad \text{transformation matrix for } C_2$$

In matrix notation,

$$\begin{bmatrix} x' \\ y' \\ z' \end{bmatrix} = \begin{bmatrix} -1 & 0 & 0 \\ 0 & -1 & 0 \\ 0 & 0 & 1 \end{bmatrix} \begin{bmatrix} x \\ y \\ z \end{bmatrix} = \begin{bmatrix} -x \\ -y \\ z \end{bmatrix} \quad \text{or} \quad \begin{bmatrix} x' \\ y' \\ z' \end{bmatrix} = \begin{bmatrix} -x \\ -y \\ z \end{bmatrix}$$

$$\begin{bmatrix} \text{New} \\ \text{coordinates} \end{bmatrix} = \begin{bmatrix} \text{transformation} \\ \text{matrix} \end{bmatrix} \begin{bmatrix} \text{old} \\ \text{coordinates} \end{bmatrix} = \begin{bmatrix} \text{new coordinates} \\ \text{in terms of old} \end{bmatrix}$$

$\sigma_v(xz)$: Reflect a point with coordinates (x, y, z) through the xz plane.

$$\begin{aligned} x' = \text{new } x &= x \\ y' = \text{new } y &= -y \\ z' = \text{new } z &= z \end{aligned} \qquad \begin{bmatrix} 1 & 0 & 0 \\ 0 & -1 & 0 \\ 0 & 0 & 1 \end{bmatrix} \quad \text{transformation matrix for } \sigma_v(xz)$$

The matrix equation is

$$\begin{bmatrix} x' \\ y' \\ z' \end{bmatrix} = \begin{bmatrix} 1 & 0 & 0 \\ 0 & -1 & 0 \\ 0 & 0 & 1 \end{bmatrix} \begin{bmatrix} x \\ y \\ z \end{bmatrix} = \begin{bmatrix} x \\ -y \\ z \end{bmatrix} \quad \text{or} \quad \begin{bmatrix} x' \\ y' \\ z' \end{bmatrix} = \begin{bmatrix} x \\ -y \\ z \end{bmatrix}$$

[3]Some sources use yz as the plane of the molecule. The assignment of B_1 and B_2 in Section 4-3-3 is reversed with this choice.

The transformation matrices for the four symmetry operations of the group are

$$E: \begin{bmatrix} 1 & 0 & 0 \\ 0 & 1 & 0 \\ 0 & 0 & 1 \end{bmatrix} \quad C_2: \begin{bmatrix} -1 & 0 & 0 \\ 0 & -1 & 0 \\ 0 & 0 & 1 \end{bmatrix} \quad \sigma_v(xz): \begin{bmatrix} 1 & 0 & 0 \\ 0 & -1 & 0 \\ 0 & 0 & 1 \end{bmatrix} \quad \sigma_v'(yz): \begin{bmatrix} -1 & 0 & 0 \\ 0 & 1 & 0 \\ 0 & 0 & 1 \end{bmatrix}$$

EXERCISE 4-5

Verify the transformation matrices for the E and $\sigma_v'(yz)$ operations of the C_{2v} point group.

This set of matrices satisfies the properties of a mathematical **group**. We call this a **matrix representation** of the C_{2v} point group. This representation is a set of matrices, each corresponding to an operation in the group; these matrices combine in the same way as the operations themselves. For example, multiplying two of the matrices is equivalent to carrying out the two corresponding operations and results in a matrix that corresponds to the resulting operation (the operations are carried out right to left, so $C_2 \times \sigma_v$ means σ_v followed by C_2):

$$C_2 \times \sigma_v(xz) = \begin{bmatrix} -1 & 0 & 0 \\ 0 & -1 & 0 \\ 0 & 0 & 1 \end{bmatrix} \begin{bmatrix} 1 & 0 & 0 \\ 0 & -1 & 0 \\ 0 & 0 & 1 \end{bmatrix} = \begin{bmatrix} -1 & 0 & 0 \\ 0 & 1 & 0 \\ 0 & 0 & 1 \end{bmatrix} = \sigma_v'(yz)$$

The matrices of the matrix representation of the C_{2v} group also describe the operations of the group shown in Figure 4-15. The C_2 and $\sigma_v'(yz)$ operations interchange H_1 and H_2, whereas E and $\sigma_v(xz)$ leave them unchanged.

Characters

The **character**, defined only for a square matrix, is the trace of the matrix, or the sum of the numbers on the diagonal from upper left to lower right. For the C_{2v} point group, the following characters are obtained from the preceding matrices:

E	C_2	$\sigma_v(xz)$	$\sigma_v'(yz)$
3	−1	1	1

We can say that this set of characters also forms a **representation**. It is an alternate shorthand version of the matrix representation. Whether in matrix or character format, this representation is called a **reducible representation**, a combination of more fundamental **irreducible representations** as described in the next section. Reducible representations are frequently designated with a capital gamma (Γ).

Reducible and irreducible representations

Each transformation matrix in the C_{2v} set above is "block diagonalized"; that is, it can be broken down into smaller matrices along the diagonal, with all other matrix elements equal to zero:

$$E: \begin{bmatrix} [1] & 0 & 0 \\ 0 & [1] & 0 \\ 0 & 0 & [1] \end{bmatrix} \quad C_2: \begin{bmatrix} [-1] & 0 & 0 \\ 0 & [-1] & 0 \\ 0 & 0 & [1] \end{bmatrix} \quad \sigma_v(xz): \begin{bmatrix} [1] & 0 & 0 \\ 0 & [-1] & 0 \\ 0 & 0 & [1] \end{bmatrix} \quad \sigma_v'(yz): \begin{bmatrix} [-1] & 0 & 0 \\ 0 & [1] & 0 \\ 0 & 0 & [1] \end{bmatrix}$$

All the nonzero elements become 1×1 matrices along the principal diagonal.

When matrices are block diagonalized in this way, the x, y, and z coordinates are also block diagonalized. As a result, the x, y, and z coordinates are independent of each other. The matrix elements in the 1,1 positions (numbered as row, column) describe the results of the symmetry operations on the x coordinate, those in the 2,2 positions describe the results of the operations on the y coordinate, and those in the 3,3 positions describe the results of the operations on the z coordinate. The four matrix elements for x form a representation of the group, those for y form a second representation, and those for z form a third representation, all shown in the following table:

Irreducible representations of the C_{2v} point group, which add to make up the reducible representation Γ

	E	C_2	$\sigma_v(xz)$	$\sigma_v'(yz)$	Coordinate Used
	1	−1	1	−1	x
	1	−1	−1	1	y
	1	1	1	1	z
Γ	3	−1	1	1	

Each row is an irreducible representation (it cannot be simplified further), and the characters of these three irreducible representations added together under each operation (column) make up the characters of the reducible representation Γ, just as the combination of all the matrices for the x, y, and z coordinates makes up the matrices of the reducible representation. For example, the sum of the three characters for x, y, and z under the C_2 operation is −1, the character for Γ under this same operation.

The set of 3×3 matrices obtained for H_2O is called a reducible representation, because it is the sum of irreducible representations (the block diagonalized 1×1 matrices), which cannot be reduced to smaller component parts. The set of characters of these matrices also forms the reducible representation Γ, for the same reason.

4-3-3 CHARACTER TABLES

Three of the representations for C_{2v}, labeled A_1, B_1, and B_2 below, have been determined so far. The fourth, called A_2, can be found by using the properties of a group described in Table 4-7. A complete set of irreducible representations for a point group is called the **character table** for that group. The character table for each point group is unique; character tables for the common point groups are included in Appendix C.

The complete character table for C_{2v} with the irreducible representations in the order commonly used, is

C_{2v}	E	C_2	$\sigma_v(xz)$	$\sigma_v'(yz)$		
A_1	1	1	1	1	z	x^2, y^2, z^2
A_2	1	1	−1	−1	R_z	xy
B_1	1	−1	1	−1	x, R_y	xz
B_2	1	−1	−1	1	y, R_x	yz

The labels used with character tables are as follows:

x, y, z	transformations of the x, y, z coordinates or combinations thereof
R_x, R_y, R_z	rotation about the x, y, and z axes
R	any symmetry operation [such as C_2 or $\sigma_v(xz)$]
χ	character of an operation
i and j	designation of different representations (such as A_1 or A_2)
h	order of the group (the total number of symmetry operations in the group)

The labels in the left column used to designate the representations will be described later in this section. Other useful terms are defined in Table 4-7.

TABLE 4-7
Properties of Characters of Irreducible Representations in Point Groups

Property	Example: C_{2v}
1. The total number of symmetry operations in the group is called the **order (h)**. To determine the order of a group, simply total the number of symmetry operations as listed in the top row of the character table.	Order $= 4$ [4 symmetry operations: E, C_2, $\sigma_v(xz)$, and $\sigma_v'(yz)$].
2. Symmetry operations are arranged in **classes**. All operations in a class have identical characters for their transformation matrices and are grouped in the same column in character tables.	Each symmetry operation is in a separate class; therefore, there are 4 columns in the character table.
3. The number of irreducible representations equals the number of classes. This means that character tables have the same number of rows and columns (they are square).	Because there are 4 classes, there must also be 4 irreducible representations—and there are.
4. The sum of the squares of the **dimensions** (characters under E) of each of the irreducible representations equals the order of the group. $$h = \sum_i [\chi_i(E)]^2$$	$1^2 + 1^2 + 1^2 + 1^2 = 4 = h$, the order of the group.
5. For any irreducible representation, the sum of the squares of the characters multiplied by the number of operations in the class (see Table 4-8 for an example), equals the order of the group. $$h = \sum_R [\chi_i(R)]^2$$	For A_2, $1^2 + 1^2 + (-1)^2 + (-1)^2 = 4 = h$. Each operation is its own class in this group.
6. Irreducible representations are **orthogonal** to each other. The sum of the products of the characters (multiplied together for each class) for any pair of irreducible representations is 0. $$\sum_R \chi_i(R)\chi_j(R) = 0, \text{ when } i \neq j$$ Taking any pair of irreducible representations, multiplying together the characters for each class and multiplying by the number of operations in the class (see Table 4-8 for an example), and adding the products gives zero.	B_1 and B_2 are orthogonal: $(1)(1)+(-1)(-1)+(1)(-1)+(-1)(1)=0$ $\quad E \qquad C_2 \qquad \sigma_v(xz) \qquad \sigma_v'(yz)$ Each operation is its own class in this group.
7. **A totally symmetric representation** is included in all groups, with characters of 1 for all operations.	C_{2v} has A_1, which has all characters $= 1$.

The A_2 representation of the C_{2v} group can now be explained. The character table has four columns; it has four classes of symmetry operations (Property 2 in Table 4-7). It must therefore have four irreducible representations (Property 3). The sum of the products of the characters of any two representations must equal zero (orthogonality, Property 6). Therefore, a product of A_1 and the unknown representation must have 1 for two of the characters and -1 for the other two. The character for the identity operation of this new representation must be 1 $[\chi(E) = 1]$ in order to have the sum of the squares of these characters equal 4 (required by Property 4). Because no two representations can be the same, A_2 must then have $\chi(E) = \chi(C_2) = 1$, and $\chi(\sigma_{xz}) = \chi(\sigma_{yz}) = -1$. This representation is also orthogonal to B_1 and B_2, as required.

Another example: $C_{3v}(NH_3)$

Full descriptions of the matrices for the operations in this group will not be given, but the characters can be found by using the properties of a group. Consider the C_3 rotation shown in Figure 4-16. Rotation of 120° results in new x' and y' as shown, which can be described in terms of the vector sums of x and y by using trigonometric functions:

$$x' = x \cos \frac{2\pi}{3} - y \sin \frac{2\pi}{3} = -\frac{1}{2}x - \frac{\sqrt{3}}{2}y$$

$$y' = x \sin \frac{2\pi}{3} + y \cos \frac{2\pi}{3} = \frac{\sqrt{3}}{2}x - \frac{1}{2}y$$

The transformation matrices for the symmetry operations shown are as follows:

$$E: \begin{bmatrix} 1 & 0 & 0 \\ 0 & 1 & 0 \\ 0 & 0 & 1 \end{bmatrix} \quad C_3: \begin{bmatrix} \cos \frac{2\pi}{3} & -\sin \frac{2\pi}{3} & 0 \\ \sin \frac{2\pi}{3} & \cos \frac{2\pi}{3} & 0 \\ 0 & 0 & 1 \end{bmatrix} = \begin{bmatrix} -\frac{1}{2} & -\frac{\sqrt{3}}{2} & 0 \\ \frac{\sqrt{3}}{2} & -\frac{1}{2} & 0 \\ 0 & 0 & 1 \end{bmatrix} \quad \sigma_{v(xz)}: \begin{bmatrix} 1 & 0 & 0 \\ 0 & -1 & 0 \\ 0 & 0 & 1 \end{bmatrix}$$

In the C_{3v} point group, $\chi(C_3{}^2) = \chi(C_3)$, which means that they are in the same class and described as $2C_3$ in the character table. In addition, the three reflections have identical characters and are in the same class, described as $3\sigma_v$.

The transformation matrices for C_3 and $C_3{}^2$ cannot be block diagonalized into 1×1 matrices because the C_3 matrix has off-diagonal entries; however, the matrices can be block diagonalized into 2×2 and 1×1 matrices, with all other matrix elements equal to zero:

$$E: \begin{bmatrix} \begin{bmatrix} 1 & 0 \\ 0 & 1 \end{bmatrix} & 0 \\ 0 & 0 & [1] \end{bmatrix} \quad C_3: \begin{bmatrix} \begin{bmatrix} -\frac{1}{2} & -\frac{\sqrt{3}}{2} \\ \frac{\sqrt{3}}{2} & -\frac{1}{2} \end{bmatrix} & 0 \\ 0 & 0 & [1] \end{bmatrix} \quad \sigma_{v(xz)}: \begin{bmatrix} \begin{bmatrix} 1 & 0 \\ 0 & -1 \end{bmatrix} & 0 \\ 0 & 0 & [1] \end{bmatrix}$$

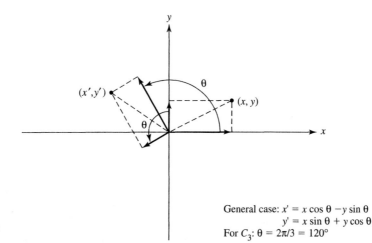

General case: $x' = x \cos \theta - y \sin \theta$
$\qquad\qquad y' = x \sin \theta + y \cos \theta$
For C_3: $\theta = 2\pi/3 = 120°$

FIGURE 4-16 Effect of Rotation on Coordinates of a Point.

The C_3 matrix must be blocked this way because the (x, y) combination is needed for the new x' and y'; the other matrices must follow the same pattern for consistency across the representation. In this case, x and y are not independent of each other.

The characters of the matrices are the sums of the numbers on the principal diagonal (from upper left to lower right). The set of 2×2 matrices has the characters corresponding to the E representation in the following character table; the set of 1×1 matrices matches the A_1 representation. The third irreducible representation, A_2, can be found by using the defining properties of a mathematical group, as in the C_{2v} example above. Table 4-8 gives the properties of the characters for the C_{3v} point group.

C_{3v}	E	$2C_3$	$3\sigma_v$		
A_1	1	1	1	z	$x^2 + y^2, z^2$
A_2	1	1	-1	R_z	
E	2	-1	0	$(x, y), (R_x, R_y)$	$(x^2 - y^2, xy), (xz, yz)$

TABLE 4-8
Properties of the Characters for the C_{3v} Point Group

Property	C_{3v} Example
1. Order	6 (6 symmetry operations)
2. Classes	3 classes: E $2C_3 (= C_3, C_3{}^2)$ $3\sigma_v (= \sigma_v, \sigma_v', \sigma_v'')$
3. Number of irreducible representations	3 (A_1, A_2, E)
4. Sum of squares of dimensions equals the order of the group	$1^2 + 1^2 + 2^2 = 6$
5. Sum of squares of characters multiplied by the number of operations in each class equals the order of the group	$\begin{array}{cccc} & E & 2C_3 & 3\sigma_v \\ \hline A_1: & 1^2 + 2(1)^2 & + 3(1)^2 & = 6 \\ A_2: & 1^2 + 2(1)^2 & + 3(-1)^2 & = 6 \\ E: & 2^2 + 2(-1)^2 & + 3(0)^2 & = 6 \end{array}$ (multiply the squares by the number of symmetry operations in each class)
6. Orthogonal representations	The sum of the products of any two representations multiplied by the number of operations in each class equals 0. Example of $A_2 \times E$: $(1)(2) + 2(1)(-1) + 3(-1)(0) = 0$
7. Totally symmetric representation	A_1, with all characters $= 1$

Additional features of character tables

1. When operations such as C_3 are in the same class, the listing in a character table is $2C_3$, indicating that the results are the same whether rotation is in a clockwise or counterclockwise direction (or, alternately, that C_3 and $C_3{}^2$ give the same result). In either case, this is equivalent to two columns in the table being shown as one. Similar notation is used for multiple reflections.

2. When necessary, the C_2 axes perpendicular to the principal axis (in a D group) are designated with primes; a single prime indicates that the axis passes through several atoms of the molecule, whereas a double prime indicates that it passes between the atoms.

3. When the mirror plane is perpendicular to the principal axis, or horizontal, the reflection is called σ_h. Other planes are labeled σ_v or σ_d; see the character tables in Appendix C.

4. The expressions listed to the right of the characters indicate the symmetry of mathematical functions of the coordinates x, y, and z and of rotation about the axes (R_x, R_y, R_z). These can be used to find the orbitals that match the representation. For example, x with positive and negative directions matches the p_x orbital with positive and negative lobes in the quadrants in the xy plane, and the product xy with alternating signs on the quadrants matches lobes of the d_{xy} orbital, as in Figure 4-17. In all cases, the totally symmetric s orbital matches the first representation in the group, one of the A set. The rotational functions are used to describe the rotational motions of the molecule. Rotation and other motions of the water molecule are discussed in Section 4-4-2.

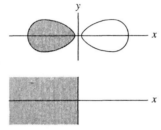

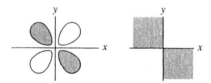

FIGURE 4-17 Orbitals and Representations.

p_x orbitals have the same symmetry as x (positive in half the quadrants, negative in the other half).

d_{xy} orbitals have the same symmetry as the function xy (sign of the function in the four quadrants).

In the C_{3v} example described previously the x and y coordinates appeared together in the E irreducible representation. The notation for this is to group them as (x, y) in this section of the table. This means that x and y together have the same symmetry properties as the E irreducible representation. Consequently, the p_x and p_y orbitals together have the same symmetry as the E irreducible representation in this point group.

5. Matching the symmetry operations of a molecule with those listed in the top row of the character table will confirm any point group assignment.

6. Irreducible representations are assigned labels according to the following rules, in which symmetric means a character of 1 and antisymmetric a character of -1 (see the character tables in Appendix C for examples).

   a. Letters are assigned according to the dimension of the irreducible representation (the character for the identity operation).

Dimension	Symmetry Label	
1	A	if the representation is symmetric to the principal rotation operation $(\chi(C_n) = 1)$.
	B	if it is antisymmetric $(\chi(C_n) = -1)$.
2	E	
3	T	

b. Subscript 1 designates a representation symmetric to a C_2 rotation perpendicular to the principal axis, and subscript 2 designates a representation antisymmetric to the C_2. If there are no perpendicular C_2 axes, 1 designates a representation symmetric to a vertical plane, and 2 designates a representation antisymmetric to a vertical plane.

c. Subscript g (gerade) designates symmetric to inversion, and subscript u (ungerade) designates antisymmetric to inversion.

d. Single primes are symmetric to σ_h and double primes are antisymmetric to σ_h when a distinction between representations is needed ($C_{3h}, C_{5h}, D_{3h}, D_{5h}$).

4-4
EXAMPLES AND APPLICATIONS OF SYMMETRY

4-4-1 CHIRALITY

Many molecules are not superimposable on their mirror image. Such molecules, labeled **chiral** or **dissymmetric**, may have important chemical properties as a consequence of this nonsuperimposability. An example of a chiral organic molecule is CBrClFI, and many examples of chiral objects can also be found on the macroscopic scale, as in Figure 4-18.

Chiral objects are termed dissymmetric. This term does not imply that these objects necessarily have *no* symmetry. For example, the propellers shown in Figure 4-18 each have a C_3 axis, yet they are nonsuperimposable (if both were spun in a clockwise direction, they would move an airplane in opposite directions!). In general, we can say that a molecule or some other object is chiral if it has no symmetry operations (other than E) or if it has *only proper rotation axes*.

EXERCISE 4-6

Which point groups are possible for chiral molecules? (Hint: Refer as necessary to the character tables in Appendix C.)

Air blowing past the stationary propellers in Figure 4-18 will be rotated in either a clockwise or counterclockwise direction. By the same token, plane-polarized light will be rotated on passing through chiral molecules (Figure 4-19); clockwise rotation is designated **dextrorotatory**, and counterclockwise rotation **levorotatory**. The ability of chiral molecules to rotate plane-polarized light is termed **optical activity** and may be measured experimentally.

Many coordination compounds are chiral and thus exhibit optical activity if they can be resolved into the two isomers. One of these is $[Ru(NH_2CH_2CH_2NH_2)_3]^{2+}$, with

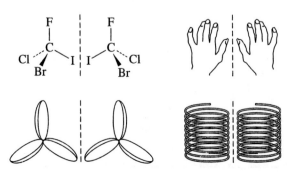

FIGURE 4-18 A Chiral Molecule and Other Chiral Objects.

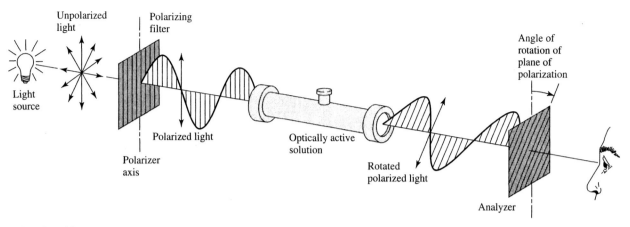

FIGURE 4-19 Rotation of Plane-Polarized Light.

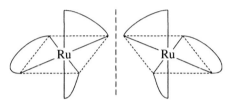

FIGURE 4-20 Chiral Isomers of $[Ru(NH_2CH_2CH_2NH_2)_3]^{2+}$.

D_3 symmetry (Figure 4-20). Mirror images of this molecule look much like left- and right-handed three-bladed propellers. Further examples will be discussed in Chapter 9.

4-4-2 MOLECULAR VIBRATIONS

Symmetry can be helpful in determining the modes of vibration of molecules. Vibrational modes of water and the stretching modes of CO in carbonyl complexes are examples that can be treated quite simply, as described in the following pages. Other molecules can be studied using the same methods.

Water (C_{2v} symmetry)

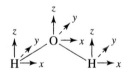

FIGURE 4-21 A Set of Axes for the Water Molecule.

Because the study of vibrations is the study of motion of the individual atoms in a molecule, we must first attach a set of x, y, and z coordinates to each atom. For convenience, we assign the z axes parallel to the C_2 axis of the molecule, the x axes in the plane of the molecule, and the y axes perpendicular to the plane (Figure 4-21). Each atom can move in all three directions, so a total of nine transformations (motion of each atom in the x, y, and z directions) must be considered. For N atoms in a molecule, there are $3N$ total motions, known as **degrees of freedom**. Degrees of freedom for different geometries are summarized in Table 4-9. Because water has three atoms, there must be nine different motions.

We will use transformation matrices to determine the symmetry of all nine motions and then assign them to translation, rotation, and vibration. Fortunately, it is only necessary to determine the characters of the transformation matrices, not the individual matrix elements.

In this case, the initial axes make a column matrix with nine elements, and each transformation matrix is 9×9. A nonzero entry appears along the diagonal of the matrix only for an atom that does not change position. If the atom changes position during the symmetry operation, a 0 is entered. If the atom remains in its original location and

TABLE 4-9
Degrees of Freedom

Number of Atoms	Total Degrees of Freedom	Translational Modes	Rotational Modes	Vibrational Modes
N (linear)	$3N$	3	2	$3N - 5$
3 (HCN)	9	3	2	4
N (nonlinear)	$3N$	3	3	$3N - 6$
3 (H_2O)	9	3	3	3

the vector direction is unchanged, a 1 is entered. If the atom remains but the vector direction is reversed, a -1 is entered. (Because all the operations change vector directions by 0° or 180° in the C_{2v} point group, these are the only possibilities.) When all nine vectors are summed, the character of the reducible representation Γ is obtained. The full 9×9 matrix for C_2 is shown as an example; note that only the diagonal entries are used in finding the character.

$$
O\left\{\begin{bmatrix} x' \\ y' \\ z' \end{bmatrix}\right. \\
H_a\left\{\begin{bmatrix} x' \\ y' \\ z' \end{bmatrix}\right. \\
H_b\left\{\begin{bmatrix} x' \\ y' \\ z' \end{bmatrix}\right. = \begin{bmatrix} -1 & 0 & 0 & 0 & 0 & 0 & 0 & 0 & 0 \\ 0 & -1 & 0 & 0 & 0 & 0 & 0 & 0 & 0 \\ 0 & 0 & 1 & 0 & 0 & 0 & 0 & 0 & 0 \\ 0 & 0 & 0 & 0 & 0 & 0 & -1 & 0 & 0 \\ 0 & 0 & 0 & 0 & 0 & 0 & 0 & -1 & 0 \\ 0 & 0 & 0 & 0 & 0 & 0 & 0 & 0 & 1 \\ 0 & 0 & 0 & -1 & 0 & 0 & 0 & 0 & 0 \\ 0 & 0 & 0 & 0 & -1 & 0 & 0 & 0 & 0 \\ 0 & 0 & 0 & 0 & 0 & 1 & 0 & 0 & 0 \end{bmatrix} \begin{bmatrix} x \\ y \\ z \\ x \\ y \\ z \\ x \\ y \\ z \end{bmatrix} \left.\begin{matrix} \ \\ \ \\ \ \end{matrix}\right\}O \left.\begin{matrix} \ \\ \ \\ \ \end{matrix}\right\}H_a \left.\begin{matrix} \ \\ \ \\ \ \end{matrix}\right\}H_b
$$

The H_a and H_b entries are not on the principal diagonal because H_a and H_b exchange with each other in a C_2 rotation, and $x'(H_a) = -x(H_b)$, $y'(H_a) = -y(H_b)$, and $z'(H_a) = z(H_b)$. Only the oxygen atom contributes to the character for this operation, for a total of -1.

The other entries for Γ can also be found without writing out the matrices, as follows:

E: All nine vectors are unchanged in the identity operation, so the character is 9.

C_2:   The hydrogen atoms change position in a C_2 rotation, so all their vectors have zero contribution to the character. The oxygen atom vectors in the x and y directions are reversed, each contributing -1, and in the z direction they remain the same, contributing 1, for a total of -1. [The sum of the principal diagonal $= \chi(C_2) = (-1) + (-1) + (1) = -1$.]

$\sigma_v(xz)$:   Reflection in the plane of the molecule changes the direction of all the y vectors and leaves the x and z vectors unchanged, for a total of $3 - 3 + 3 = 3$.

$\sigma_v'(yz)$:   Finally, reflection perpendicular to the plane of the molecule changes the position of the hydrogens so their contribution is zero; the x vector on the oxygen changes direction and the y and z vectors are unchanged, for a total of 1.

Because all nine direction vectors are included in this representation, it represents all the motions of the molecule, three translations, three rotations, and (by difference) three vibrations. The characters of the reducible representation Γ are shown as the last row below the irreducible representations in the C_{2v} character table.

C_{2v}	E	C_2	$\sigma_v(xz)$	$\sigma_v'(yz)$		
A_1	1	1	1	1	z	x^2, y^2, z^2
A_2	1	1	−1	−1	R_z	xy
B_1	1	−1	1	−1	x, R_y	xz
B_2	1	−1	−1	1	y, R_x	yz
Γ	9	−1	3	1		

Reducing representations to irreducible representations

The next step is to separate this representation into its component irreducible representations. This requires another property of groups. The number of times that any irreducible representation appears in a reducible representation is equal to the sum of the products of the characters of the reducible and irreducible representations taken one operation at a time, divided by the order of the group. This may be expressed in equation form, with the sum taken over all symmetry operations of the group.[4]

$$\begin{pmatrix} \text{Number of irreducible} \\ \text{representations of} \\ \text{a given type} \end{pmatrix} = \frac{1}{\text{order}} \sum_R \left[\begin{pmatrix} \text{number} \\ \text{of operations} \\ \text{in the class} \end{pmatrix} \times \begin{pmatrix} \text{character of} \\ \text{reducible} \\ \text{representation} \end{pmatrix} \times \begin{pmatrix} \text{character of} \\ \text{irreducible} \\ \text{representation} \end{pmatrix} \right]$$

In the water example, the order of C_{2v} is 4, with one operation in each class $(E, C_2, \sigma_v, \sigma_v')$. The results are then

$$n_{A_1} = \frac{1}{4}[(9)(1) + (-1)(1) + (3)(1) + (1)(1)] = 3$$

$$n_{A_2} = \frac{1}{4}[(9)(1) + (-1)(1) + (3)(-1) + (1)(-1)] = 1$$

$$n_{B_1} = \frac{1}{4}[(9)(1) + (-1)(-1) + (3)(1) + (1)(-1)] = 3$$

$$n_{B_2} = \frac{1}{4}[(9)(1) + (-1)(-1) + (3)(-1) + (1)(1)] = 2$$

The reducible representation for all motions of the water molecule is therefore reduced to $3A_1 + A_2 + 3B_1 + 2B_2$.

Examination of the columns on the far right in the character table shows that translation along the x, y, and z directions is $A_1 + B_1 + B_2$ (translation is motion along the x, y, and z directions, so it transforms in the same way as the three axes) and that rotation in the three directions (R_x, R_y, R_z) is $A_2 + B_1 + B_2$. Subtracting these from the total above leaves $2A_1 + B_1$, the three vibrational modes, as shown in Table 4-10. The number of vibrational modes equals $3N − 6$, as described earlier. Two of the modes are totally symmetric (A_1) and do not change the symmetry of the molecule, but one is antisymmetric to C_2 rotation and to reflection perpendicular to the plane of the molecule (B_1). These modes are illustrated as symmetric stretch, symmetric bend, and antisymmetric stretch in Table 4-11.

---

[4]This procedure should yield an integer for the number of irreducible representations of each type; obtaining a fraction in this step indicates a calculation error.

TABLE 4-10
Symmetry of Molecular Motions of Water

All Motions	Translation (x, y, z)	Rotation (R_x, R_y, R_z)	Vibration (Remaining Modes)
$3A_1$	A_1		$2A_1$
A_2		A_2	
$3B_1$	B_1	B_1	B_1
$2B_2$	B_2	B_2	

TABLE 4-11
The Vibrational Modes of Water

A_1

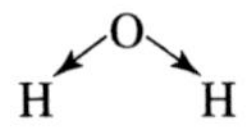

Symmetric stretch: change in dipole moment; more distance between positive hydrogens and negative oxygen
IR active

B_1

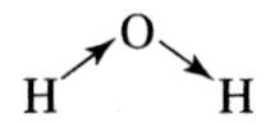

Antisymmetric stretch: change in dipole moment; change in distances between positive hydrogens and negative oxygen
IR active

A_1

Symmetric bend: change in dipole moment; angle between H—O vectors changes
IR active

A molecular vibration is infrared active (has an infrared absorption) only if it results in a change in the dipole moment of the molecule. The three vibrations of water can be analyzed this way to determine their infrared behavior. In fact, the oxygen atom also moves. Its motion is opposite that of the hydrogens and is very small, because its mass is so much larger than that of the hydrogen atoms. The center of mass of the molecule does not move in vibrations.

Group theory can give us the same information (and can account for the more complicated cases as well; in fact, group theory in principle can account for *all* vibrational modes of a molecule). In group theory terms, a vibrational mode is active in the infrared *if it corresponds to an irreducible representation that has the same symmetry (or transforms) as the Cartesian coordinates x, y, or z,* because a vibrational motion that shifts the center of charge of the molecule in any of the *x, y,* or *z* directions results in a change in dipole moment. Otherwise, the vibrational mode is not infrared active.

EXAMPLES

Reduce the following representations to their irreducible representations in the point group indicated (refer to the character tables in Appendix C):

C_{2h}	E	C_2	i	σ_h
Γ	4	0	2	2

Solution:

$$n_{A_g} = \frac{1}{4}[(4)(1) + (0)(1) + (2)(1) + (2)(1)] = 2$$

$$n_{B_g} = \frac{1}{4}[(4)(1) + (0)(-1) + (2)(1) + (2)(-1)] = 1$$

$$n_{A_u} = \frac{1}{4}[(4)(1) + (0)(1) + (2)(-1) + (2)(-1)] = 0$$

$$n_{B_u} = \frac{1}{4}[(4)(1) + (0)(-1) + (2)(-1) + (2)(1)] = 1$$

Therefore, $\Gamma = 2A_g + B_g + B_u$.

C_{3v}	E	$2C_3$	$3\sigma_v$
Γ	6	3	-2

Solution:

$$n_{A_1} = \frac{1}{6}[(6)(1) + (2)(3)(1) + (3)(-2)(1)] = 1$$

$$n_{A_2} = \frac{1}{6}[(6)(1) + (2)(3)(1) + (3)(-2)(-1)] = 3$$

$$n_E = \frac{1}{6}[(6)(2) + (2)(3)(-1) + (3)(-2)(0)] = 1$$

Therefore, $\Gamma = A_1 + 3A_2 + E$.

Be sure to include the number of symmetry operations in a class (column) of the character table. This means that the second term in the C_{3v} calculation must be multiplied by 2 ($2C_3$; there are two operations in this class), and the third term must be multiplied by 3, as shown.

EXERCISE 4-7

Reduce the following representations to their irreducible representations in the point groups indicated:

T_d	E	$8C_3$	$3C_2$	$6S_4$	$6\sigma_d$
Γ_1	4	1	0	0	2

D_{2d}	E	$2S_4$	C_2	$2C_2'$	$2\sigma_d$
Γ_2	4	0	0	2	0

C_{4v}	E	$2C_4$	C_2	$2\sigma_v$	$2\sigma_d$
Γ_3	7	-1	-1	-1	-1

EXERCISE 4-8

Analysis of the x, y, and z coordinates of each atom in NH_3 gives the following representation:

C_{3v}	E	$2C_3$	$3\sigma_v$
Γ	12	0	2

a. Reduce Γ to its irreducible representations.

b. Classify the irreducible representations into translational, rotational, and vibrational modes.

c. Show that the total number of degrees of freedom $= 3N$.

d. Which vibrational modes are infrared active?

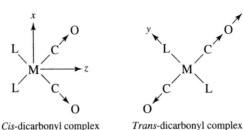

FIGURE 4-22 Carbonyl Stretching Vibrations of *cis-* and *trans-*Dicarbonyl Square Planar Complexes.

Cis-dicarbonyl complex   *Trans*-dicarbonyl complex

Selected vibrational modes

It is often useful to consider a particular type of vibrational mode for a compound. For example, useful information often can be obtained from the C—O stretching bands in infrared spectra of metal complexes containing CO (carbonyl) ligands. The following example of *cis-* and *trans-*dicarbonyl square planar complexes shows the procedure. For these complexes,[5] a simple IR spectrum can distinguish whether a sample is *cis-* or *trans*-$ML_2(CO)_2$; the number of C—O stretching bands is determined by the geometry of the complex (Figure 4-22).

***cis*-$ML_2(CO)_2$, point group C_{2v}.**   The principal axis (C_2) is the z axis, with the xz plane assigned as the plane of the molecule. Possible C—O stretching motions are shown by arrows in Figure 4-23; either an increase or decrease in the C—O distance is possible. These vectors are used to create the reducible representation below using the symmetry operations of the C_{2v} point group. A C—O bond will transform with a character of 1 *if it remains unchanged* by the symmetry operations, and with a character of 0 *if it is changed*. These operations and their characters are shown in Figure 4-23. Both

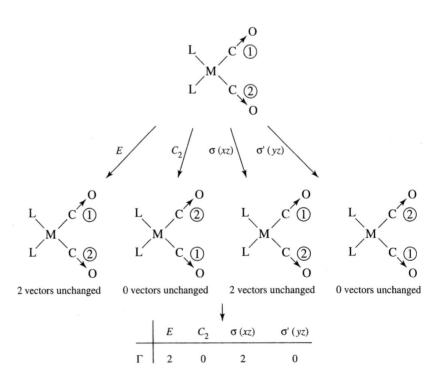

2 vectors unchanged   0 vectors unchanged   2 vectors unchanged   0 vectors unchanged

	E	C_2	$\sigma (xz)$	$\sigma' (yz)$
Γ	2	0	2	0

FIGURE 4-23 Symmetry Operations and Characters for *cis*-$ML_2(CO)_2$.

[5]M represents any metal and L any ligand other than CO in these formulas.

stretches are unchanged in the identity operation and in the reflection through the plane of the molecule, so each contributes 1 to the character, for a total of 2 for each operation. Both vectors move to new locations on rotation or reflection perpendicular to the plane of the molecule, so these two characters are 0.

The reducible representation Γ reduces to $A_1 + B_1$:

C_{2v}	E	C_2	$\sigma_v(xz)$	$\sigma_v'(yz)$		
Γ	2	0	2	0		
A_1	1	1	1	1	z	x^2, y^2, z^2
B_1	1	-1	1	-1	x, R_y	xz

A_1 is an appropriate irreducible representation for an IR-active band, because it transforms as (has the symmetry of) the Cartesian coordinate z. Furthermore, the vibrational mode corresponding to B_1 should be IR active, because it transforms as the Cartesian coordinate x.

In summary:

There are two vibrational modes for C—O stretching, one having A_1 symmetry and one B_1 symmetry. Both modes are IR active, and we therefore expect to see two C—O stretches in the IR. This assumes that the C—O stretches are not sufficiently similar in energy to overlap in the infrared spectrum.

trans-ML$_2$(CO)$_2$, point group D_{2h}. The principal axis, C_2, is again chosen as the z axis, which this time makes the plane of the molecule the xy plane. Using the symmetry operation of the D_{2h} point group, we obtain a reducible representation for C—O stretches that reduces to $A_g + B_{3u}$:

D_{2h}	E	$C_2(z)$	$C_2(y)$	$C_2(x)$	i	$\sigma(xy)$	$\sigma(xz)$	$\sigma(yz)$	
Γ	2	0	0	2	0	2	2	0	
A_g	1	1	1	1	1	1	1	1	x^2, y^2, z^2
B_{3u}	1	-1	-1	1	-1	1	1	-1	x

The vibrational mode of A_g symmetry is not IR active, because it does not have the same symmetry as a Cartesian coordinate x, y, or z (this is the IR-inactive symmetric stretch). The mode of symmetry B_{3u}, on the other hand, is IR active, because it has the same symmetry as x.

In summary:

There are two vibrational modes for C—O stretching, one having the same symmetry as A_g, and one the same symmetry as B_{3u}. The A_g mode is IR inactive (does not have the symmetry of x, y, or z); the B_{3u} mode is IR active (has the symmetry of x). We therefore expect to see one C—O stretch in the IR.

It is therefore possible to distinguish _cis_- and _trans_-ML$_2$(CO)$_2$ by taking an IR spectrum. If one C—O stretching band appears, the molecule is _trans_; if two bands appear, the molecule is _cis_. A significant distinction can be made by a very simple measurement.

EXAMPLE

Determine the number of IR-active CO stretching modes for *fac*-$Mo(CO)_3(NCCH_3)_3$, as shown in the diagram.

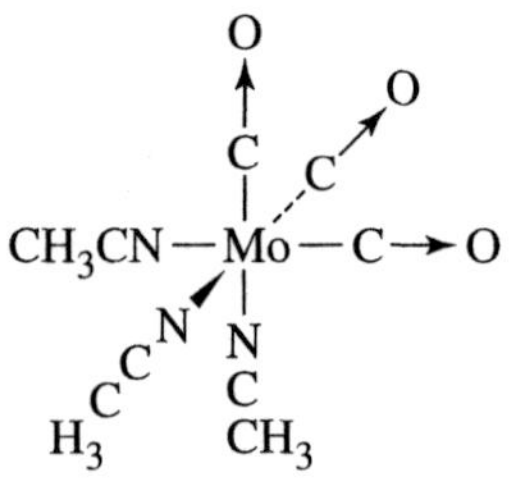

This molecule has C_{3v} symmetry. The operations to be considered are E, C_3, and σ_v. E leaves the three bond vectors unchanged, giving a character of 3. C_3 moves all three vectors, giving a character of 0. Each σ_v plane passes through one of the CO groups, leaving it unchanged, while interchanging the other two. The resulting character is 1.

The representation to be reduced, therefore, is

E	$2C_3$	$3\sigma_v$
3	0	1

This reduces to $A_1 + E$. A_1 has the same symmetry as the Cartesian coordinate z and is therefore IR active. E has the same symmetry as the x and y coordinates together and is also IR active. It represents a degenerate pair of vibrations, which appear as one absorption band.

EXERCISE 4-9

Determine the number of IR-active $C\!-\!O$ stretching modes for $Mn(CO)_5Cl$.

GENERAL REFERENCES

There are several helpful books on this subject. Good examples are F. A. Cotton, *Chemical Applications of Group Theory*, 3rd ed., John Wiley & Sons, New York, 1990; S. F. A. Kettle, *Symmetry and Structure (Readable Group Theory for Chemists)*, 2nd ed., John Wiley & Sons, New York, 1995; and I. Hargittai and M. Hargittai, *Symmetry Through the Eyes of a Chemist*, 2nd ed., Plenum Press, New York, 1995. The latter two also provide information on space groups used in solid state symmetry, and all give relatively gentle introductions to the mathematics of the subject.

PROBLEMS

4-1 Determine the point groups for
  a. Ethane (staggered conformation)
  b. Ethane (eclipsed conformation)
  c. Chloroethane (staggered conformation)
  d. 1,2-Dichloroethane (staggered *anti* conformation)

4-2 Determine the point groups for

  a. Ethylene

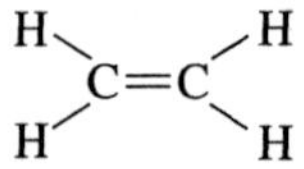

  b. Chloroethylene
  c. The possible isomers of dichloroethylene

4-3 Determine the point groups for
  a. Acetylene
  b. $H\!-\!C\!\equiv\!C\!-\!F$
  c. $H\!-\!C\!\equiv\!C\!-\!CH_3$
  d. $H\!-\!C\!\equiv\!C\!-\!CH_2Cl$
  e. $H\!-\!C\!\equiv\!C\!-\!Ph$ (Ph = phenyl)

4-4   Determine the point groups for

    a. Naphthalene

    b. 1,8-Dichloronaphthalene

    c. 1,5-Dichloronaphthalene

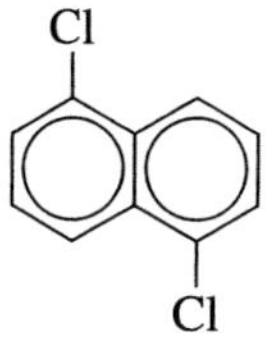

    d. 1,2-Dichloronaphthalene

4-5   Determine the point groups for

    a. 1,1′-Dichloroferrocene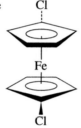

    b. Dibenzenechromium (eclipsed conformation)

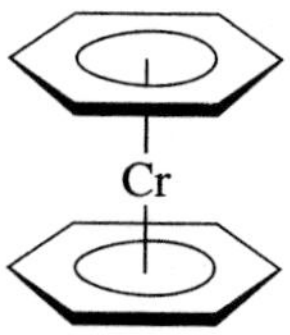

    c.

    d. H_3O^+

    e. O_2F_2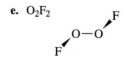

    f. Formaldehyde, H_2CO

    g. S_8 (puckered ring)

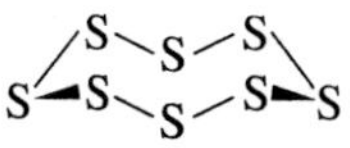

h. Borazine (planar)

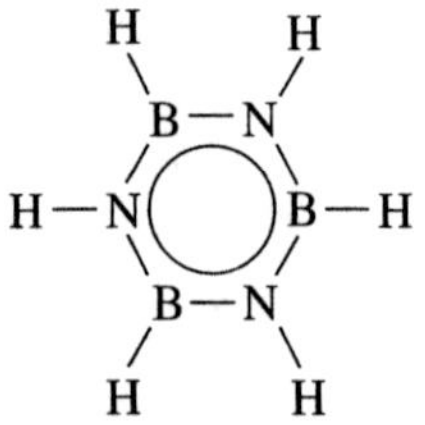

i. $[Cr(C_2O_4)_3]^{3-}$

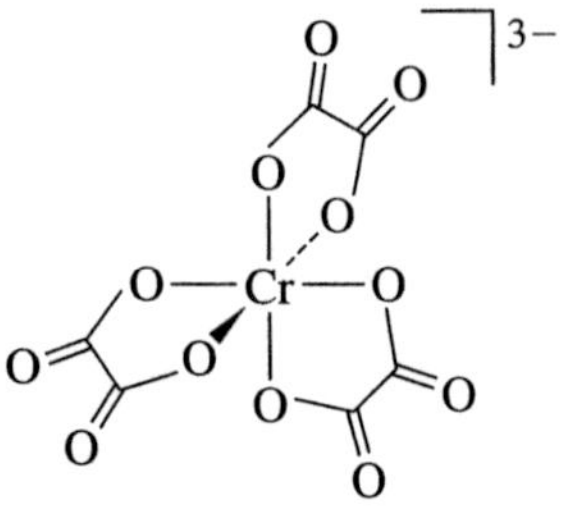

j. A tennis ball (ignoring the label, but including the pattern on the surface)

4-6 Determine the point group for
a. Cyclohexane (chair conformation)
b. Tetrachloroallene $Cl_2C{=}C{=}CCl_2$
c. $SO_4{}^{2-}$
d. A snowflake
e. Diborane

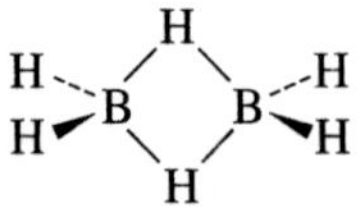

f. The possible isomers of tribromobenzene
g. A tetrahedron inscribed in a cube (alternate corners of the cube are also corners of the tetrahedron).
h. B_3H_8

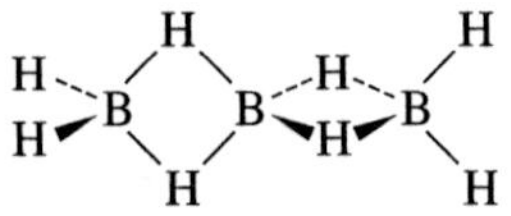

4-7 Determine the point group for
a. A sheet of typing paper
b. An Erlenmeyer flask (no label)
c. A screw
d. The number 96
e. Five examples of objects from everyday life; select items from five different point groups.
f. A pair of eyeglasses (assuming lenses of equal strength)
g. A five-pointed star
h. A fork (assuming no decoration)
i. Captain Ahab, who lost a leg to Moby Dick
j. A metal washer

4-8 Determine the point groups of the molecules in the following end-of-chapter problems from Chapter 3:
a. Problem 3-12
b. Problem 3-16

4-9 Determine the point groups of the molecules and ions in
a. Figure 3-8
b. Figure 3-15

4-10 Determine the point groups of the following atomic orbitals, including the signs on the orbital lobes:
a. p_x
b. d_{xy}
c. $d_{x^2-y^2}$
d. d_{z^2}

4-11 Show that a cube has the same symmetry elements as an octahedron.

4-12 For *trans*-1,2-dichloroethylene, of C_{2h} symmetry,
  a. List all the symmetry operations for this molecule.
  b. Write a set of transformation matrices that describe the effect of each symmetry operation in the C_{2h} group on a set of coordinates x, y, z for a point. (Your answer should consist of four 3×3 transformation matrices.)
  c. Using the terms along the diagonal, obtain as many irreducible representations as possible from the transformation matrices. (You should be able to obtain three irreducible representations in this way, but two will be duplicates.) You may check your results using the C_{2h} character table.
  d. Using the C_{2h} character table, verify that the irreducible representations are mutually orthogonal.

4-13 Ethylene is a molecule of D_{2h} symmetry.
  a. List all the symmetry operations of ethylene.
  b. Write a transformation matrix for each symmetry operation that describes the effect of that operation on the coordinates of a point x, y, z.
  c. Using the characters of your transformation matrices, obtain a reducible representation.
  d. Using the diagonal elements of your matrices, obtain three of the D_{2h} irreducible representations.
  e. Show that your irreducible representations are mutually orthogonal.

4-14 Using the D_{2d} character table,
  a. Determine the order of the group.
  b. Verify that the E irreducible representation is orthogonal to each of the other irreducible representations.
  c. For each of the irreducible representations, verify that the sum of the squares of the characters equals the order of the group.
  d. Reduce the following representations to their component irreducible representations:

D_{2d}	E	$2S_4$	C_2	$2C_2'$	$2\sigma_d$
Γ_1	6	0	2	2	2
Γ_2	6	4	6	2	0

4-15 Reduce the following representations to irreducible representations:

C_{3v}	E	$2C_3$	$3\sigma_v$
Γ_1	6	3	2
Γ_2	5	−1	−1

O_h	E	$8C_3$	$6C_2$	$6C_4$	$3C_2$	i	$6S_4$	$8S_6$	$3\sigma_h$	$6\sigma_d$
Γ	6	0	0	2	2	0	0	0	4	2

4-16 For D_{4h} symmetry show, using sketches, that d_{xy} orbitals have B_{2g} symmetry and that $d_{x^2-y^2}$ orbitals have B_{1g} symmetry. (Hint: you may find it useful to select a molecule that has D_{4h} symmetry as a reference for the operations of the D_{4h} point group.)

4-17 Which items in Problems 5, 6, and 7 are chiral? List three items *not* from this chapter that are chiral.

4-18 For the following molecules, determine the number of IR-active C—O stretching vibrations:

a.

b.

4-19 Using the x, y, and z coordinates for each atom in SF_6, determine the reducible representation, reduce it, classify the irreducible representations into translational, rotational, and vibrational modes, and decide which vibrational modes are infrared active.

4-20 Three isomers of $W_2Cl_4(NHEt)_2(PMe_3)_2$ have been reported. These isomers have the core structures shown below. Determine the point group of each (Reference: F. A. Cotton, E. V. Dikarev, and W.-Y. Wong, *Inorg. Chem.*, **1997**, *36*, 2670.)

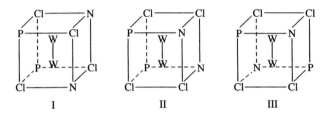

I   II   III

4-21 There is considerable evidence for the existence of protonated methane, CH_5^+. Calculations have indicated that the lowest energy form of this ion has C_s symmetry. Sketch a reasonable structure for this structure. The structure is unusual, with a type of bonding only mentioned briefly in previous chapters. (Reference: G. A. Olah and G. Rasul, *Acc. Chem. Res.*, **1997**, *30*, 245.)

4-22 The hexaazidoarsenate(V) ion, $[As(N_3)_6]^-$, has been reported as the first structurally characterized binary arsenic (V) azide species. Two views of its structure are shown below. A view with three As — N bonds pointing up and three pointing down (alternating) is shown in (a); a view down one of the N — As — N axes is shown in (b). What is its point group? (Reference: T. M. Klapötke, H. Nöth, T. Schütt, and M. Warchhold, *Angew Chem., Int. Ed.*, **2000**, *39*, 2108.)

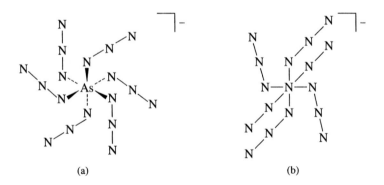

(a)   (b)

4-23 Derivatives of methane can be obtained by replacing one or more hydrogen atoms with other atoms, such as F, Cl, or Br. Suppose you had a supply of methane and the necessary chemicals and equipment to make derivatives of methane containing all possible combinations of the elements H, F, Cl, and Br. What would be the point groups of the molecules you could make? You should find 35 possible molecules, but they can be arranged into five sets for assignment of point groups.

4-24 Determine the point groups of the following molecules:
a. F_3SCCF_3, with a triple $S \equiv C$ bond

b. $C_6H_6F_2Cl_2Br_2$, a derivative of cyclohexane, in a chair conformation

c. $M_2Cl_6Br_4$, where M is a metal atom

d. $M(NH_2C_2H_4PH_2)_3$, considering the $NH_2C_2H_4PH_2$ rings as planar

4-25 Use the Internet to search for molecules with the symmetry of
a. The S_6 point group
b. The T point group
c. The I_h point group
Report the molecules, the URL of the Web site where you found them, and the search strategy you used.

CHAPTER

5

Molecular Orbitals

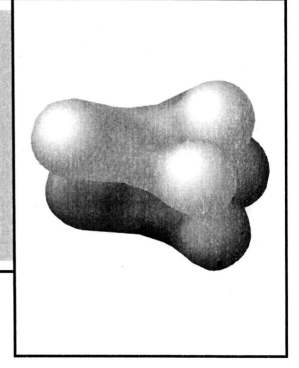

Molecular orbital theory uses the methods of group theory to describe the bonding in molecules and complements and extends the simple pictures of bonding introduced in Chapter 3. The symmetry properties and relative energies of atomic orbitals determine how they interact to form molecular orbitals. These molecular orbitals are then filled with the available electrons according to the same rules used for atomic orbitals, and the total energy of the electrons in the molecular orbitals is compared with the initial total energy of electrons in the atomic orbitals. If the total energy of the electrons in the molecular orbitals is less than in the atomic orbitals, the molecule is stable compared with the atoms; if not, the molecule is unstable and the compound does not form. We will first describe the bonding (or lack of it) in the first ten homonuclear diatomic molecules (H_2 through Ne_2) and then expand the treatment to heteronuclear diatomic molecules and to molecules having more than two atoms.

A simple pictorial approach is adequate to describe bonding in many cases and can provide clues to more complete descriptions in more difficult cases. On the other hand, it is helpful to know how a more elaborate group theoretical approach can be used, both to provide background for the simpler approach and to have it available in cases in which it is needed. In this chapter, we will describe both approaches, showing the simpler pictorial approach and developing the symmetry arguments required for some of the more complex cases.

5-1
FORMATION OF MOLECULAR ORBITALS FROM ATOMIC ORBITALS

As in the case of atomic orbitals, Schrödinger equations can be written for electrons in molecules. Approximate solutions to these molecular Schrödinger equations can be constructed from **linear combinations of the atomic orbitals (LCAO)**, the sums and differences of the atomic wave functions. For diatomic molecules such as H_2, such wave functions have the form

$$\Psi = c_a\psi_a + c_b\psi_b,$$

where Ψ is the molecular wave function, ψ_a and ψ_b are atomic wave functions, and c_a

and c_b are adjustable coefficients. The coefficients can be equal or unequal, positive or negative, depending on the individual orbitals and their energies. As the distance between two atoms is decreased, their orbitals overlap, with significant probability for electrons from both atoms in the region of overlap. As a result, **molecular orbitals** form. Electrons in bonding molecular orbitals occupy the space between the nuclei, and the electrostatic forces between the electrons and the two positive nuclei hold the atoms together.

Three conditions are essential for overlap to lead to bonding. First, the symmetry of the orbitals must be such that regions with the same sign of ψ overlap. Second, the energies of the atomic orbitals must be similar. When the energies differ by a large amount, the change in energy on formation of the molecular orbitals is small and the net reduction in energy of the electrons is too small for significant bonding. Third, the distance between the atoms must be short enough to provide good overlap of the orbitals, but not so short that repulsive forces of other electrons or the nuclei interfere. When these conditions are met, the overall energy of the electrons in the occupied molecular orbitals will be lower in energy than the overall energy of the electrons in the original atomic orbitals, and the resulting molecule has a lower total energy than the separated atoms.

5-1-1 MOLECULAR ORBITALS FROM s ORBITALS

We will consider first the combination of two s orbitals, as in H_2. For convenience, we label the atoms of a diatomic molecule a and b, so the atomic orbital wave functions are $\psi(1s_a)$ and $\psi(1s_b)$. We can visualize the two atoms moving closer to each other until the electron clouds overlap and merge into larger molecular electron clouds. The resulting molecular orbitals are linear combinations of the atomic orbitals, the sum of the two orbitals and the difference between them:

In general terms   For H_2

$$\Psi(\sigma) = N[c_a\psi(1s_a) + c_b\psi(1s_b)] = \frac{1}{\sqrt{2}}[\psi(1s_a) + \psi(1s_b)] \qquad (H_a + H_b)$$

and

$$\Psi(\sigma^*) = N[c_a\psi(1s_a) - c_b\psi(1s_b)] = \frac{1}{\sqrt{2}}[\psi(1s_a) - \psi(1s_b)] \qquad (H_a - H_b)$$

N is the normalizing factor (so $\int \Psi\Psi^* \, d\tau = 1$), and c_a and c_b are adjustable coefficients. In this case, the two atomic orbitals are identical and the coefficients are nearly identical as well.[1] These orbitals are depicted in Figure 5-1. In this diagram, as in all the orbital diagrams in this book (such as Table 2-3 and Figure 2-6), the signs of orbital lobes are indicated by shading. Light and dark lobes indicate opposite signs of Ψ. The choice of positive and negative for specific atomic orbitals is arbitrary; what is important is how they fit together to form molecular orbitals. In the diagrams on the right side in the figure, light and dark shading show opposite signs of the wave function.

---

[1]More precise calculations show that the coefficients of the σ^* orbital are slightly larger than for the σ orbital, but this difference is usually ignored in the simple approach we use. For identical atoms, we will use $c_a = c_b = 1$ and $N = 1/\sqrt{2}$. The difference in coefficients for the σ and σ^* orbitals also results in a larger energy change (increase) from atomic to the σ^* molecular orbitals than for the σ orbitals (decrease).

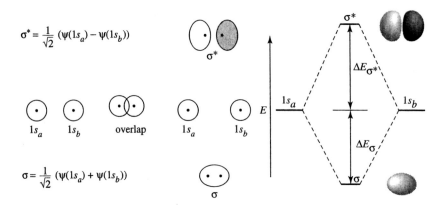

$$\sigma^* = \frac{1}{\sqrt{2}}\left(\psi(1s_a) - \psi(1s_b)\right)$$

$$\sigma = \frac{1}{\sqrt{2}}\left(\psi(1s_a) + \psi(1s_b)\right)$$

FIGURE 5-1 Molecular Orbitals from Hydrogen $1s$ Orbitals.

Because the σ molecular orbital is the sum of the two atomic orbitals, $\frac{1}{\sqrt{2}}[\psi(1s_a) + \psi(1s_b)]$, and results in an increased concentration of electrons between the two nuclei where both atomic wave functions contribute, it is a **bonding molecular orbital** and has a lower energy than the starting atomic orbitals. The σ^* molecular orbital is the difference of the two atomic orbitals, $\frac{1}{\sqrt{2}}[\psi(1s_a) - \psi(1s_b)]$. It has a node with zero electron density between the nuclei caused by cancellation of the two wave functions and has a higher energy; it is therefore called an **antibonding orbital**. Electrons in bonding orbitals are concentrated between the nuclei and attract the nuclei and hold them together. Antibonding orbitals have one or more nodes between the nuclei; electrons in these orbitals cause a mutual repulsion between the atoms. The difference in energy between an antibonding orbital and the initial atomic orbitals is slightly larger than the same difference between a bonding orbital and the initial atomic orbitals. **Nonbonding orbitals** are also possible. The energy of a nonbonding orbital is essentially that of an atomic orbital, either because the orbital on one atom has a symmetry that does not match any orbitals on the other atom, or the energy of the molecular orbital matches that of the atomic orbital by coincidence.

The σ (sigma) notation indicates orbitals that are symmetric to rotation about the line connecting the nuclei:

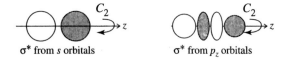

An asterisk is frequently used to indicate antibonding orbitals, the orbitals of higher energy. Because the bonding, nonbonding, or antibonding nature of a molecular orbital is sometimes uncertain, the asterisk notation will be used only in the simpler cases in which the bonding and antibonding characters are clear.

The pattern described for H_2 is the usual model for combining two orbitals: two atomic orbitals combine to form two molecular orbitals, one bonding orbital with a lower energy and one antibonding orbital with a higher energy. Regardless of the number of orbitals, the unvarying rule is that the number of resulting molecular orbitals is the same as the initial number of atomic orbitals in the atoms.

5-1-2 MOLECULAR ORBITALS FROM *p* ORBITALS

Molecular orbitals formed from *p* orbitals are more complex because of the symmetry of the orbitals. The algebraic sign of the wave function must be included when interactions between the orbitals are considered. When two orbitals overlap and the overlapping regions have the same sign, the sum of the two orbitals has an increased electron probability in the overlap region. When two regions of opposite sign overlap, the combination has a decreased electron probability in the overlap region. Figure 5-1 shows this effect for the sum and difference of the 1*s* orbitals of H_2; similar effects result from overlapping lobes of *p* orbitals with their alternating signs. The interactions of *p* orbitals are shown in Figure 5-2. For convenience, we will choose a common *z* axis connecting the nuclei. Once the axes are set for a particular molecule, they do not change.

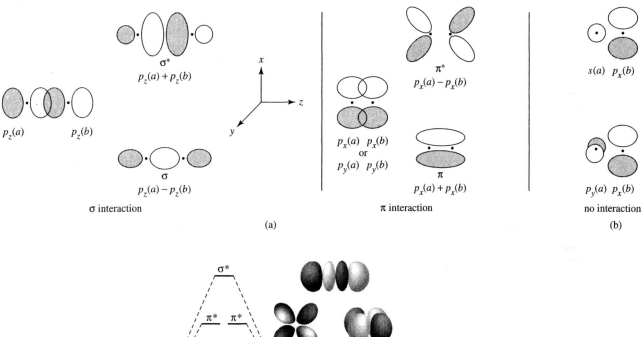

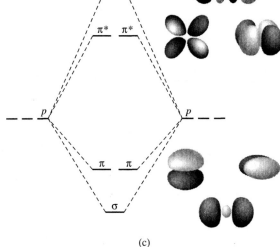

FIGURE 5-2 Interactions of *p* Orbitals. (a) Formation of molecular orbitals. (b) Orbitals that do not form molecular orbitals. (c) Energy level diagram.

When we draw the z axes for the two atoms pointing in the same direction,[2] the p_z orbitals subtract to form σ and add to form σ^* orbitals, both of which are symmetric to rotation about the z axis and with nodes perpendicular to the line that connects the nuclei. Interactions between p_x and p_y orbitals lead to π and π^* orbitals, as shown. The π (pi) notation indicates a change in sign with C_2 rotation about the bond axis:

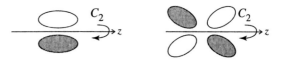

As in the case of the s orbitals, the overlap of two regions with the same sign leads to an increased concentration of electrons, and the overlap of two regions of opposite sign leads to a node of zero electron density. In addition, the nodes of the atomic orbitals become the nodes of the resulting molecular orbitals. In the π^* antibonding case, four lobes result that are similar in appearance to an expanded d orbital (Figure 5-2(c)).

The p_x, p_y, and p_z orbital pairs need to be considered separately. Because the z axis was chosen as the internuclear axis, the orbitals derived from the p_z orbitals are symmetric to rotation around the bond axis and are labeled σ and σ^* for the bonding and antibonding orbitals, respectively. Similar combinations of the p_y orbitals form orbitals whose wave functions change sign with C_2 rotation about the bond axis; they are labeled π and π^* for the bonding and antibonding orbitals, respectively. In the same way, the p_x orbitals also form π and π^* orbitals.

When orbitals overlap equally with both the same and opposite signs, as in the $s + p_x$ example in Figure 5-2(b), the bonding and antibonding effects cancel and no molecular orbital results. Another way to describe this is that the symmetry properties of the orbitals do not match and no combination is possible. If the symmetry of an atomic orbital does not match *any* orbital of the other atom, it is called a nonbonding orbital. Homonuclear diatomic molecules have only bonding and antibonding molecular orbitals; nonbonding orbitals are described further in Sections 5-1-4, 5-2-2, and 5-4-3.

5-1-3 MOLECULAR ORBITALS FROM *d* ORBITALS

In the heavier elements, particularly the transition metals, d orbitals can be involved in bonding in a similar way. Figure 5-3 shows the possible combinations. When the z axes are collinear, two d_{z^2} orbitals can combine end on for σ bonding. The d_{xz} and d_{yz} orbitals form π orbitals. When atomic orbitals meet from two parallel planes and combine side to side, as do the $d_{x^2-y^2}$ and d_{xy} orbitals with collinear z axes, they form δ (delta) orbitals. (The δ notation indicates sign changes on C_4 rotation about the bond axis.) Sigma orbitals have no nodes that include the line of centers of the atoms, pi orbitals have one node that includes the line of centers, and delta orbitals have two nodes that include the line of centers. Combinations of orbitals involving overlapping regions with opposite signs cannot form molecular orbitals; for example, p_z and d_{xz} have zero net overlap (one region with overlapping regions of the same sign and another with opposite signs).

---

[2]The choice of direction of the z axes is arbitrary. When both are positive in the same direction, ⌐⋈→ ⌐⋈→, **the difference between the p_z orbitals is the bonding combination**. When the positive z axes are chosen to point toward each other, ⌐⋈→ ←⋈⌐, the sum of the p_z orbitals is the bonding combination. We have chosen to have them positive in the same direction for consistency with our treatment of triatomic and larger molecules.

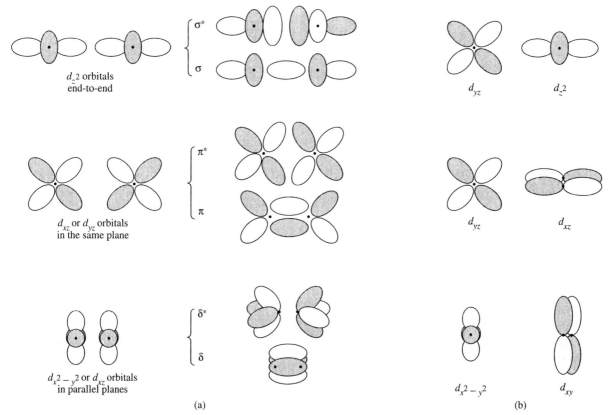

FIGURE 5-3 Interactions of *d* Orbitals. (a) Formation of molecular orbitals. (b) Orbitals that do not form molecular orbitals.

EXAMPLE

Sketch the overlap regions of the following combination of orbitals, all with collinear z axes. Classify the interactions.

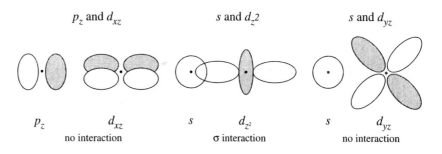

EXERCISE 5-1

Repeat the process for the preceding example for the following orbital combinations, again using collinear z axes.

$$p_x \text{ and } d_{xz} \qquad p_z \text{ and } d_{z^2} \qquad s \text{ and } d_{x^2-y^2}$$

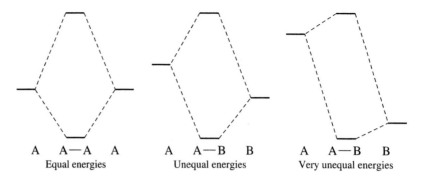

FIGURE 5-4 Energy Match and Molecular Orbital Formation.

A A—A A	A A—B B	A A—B B
Equal energies	Unequal energies	Very unequal energies

5-1-4 NONBONDING ORBITALS AND OTHER FACTORS

As mentioned previously, there can also be nonbonding molecular orbitals, whose energy is essentially that of the original atomic orbitals. These can form when there are three atomic orbitals of the same symmetry and similar energies, a situation that requires the formation of three molecular orbitals. One is a low-energy bonding orbital, one is a high-energy antibonding orbital, and one is of intermediate energy and is a nonbonding orbital. Examples will be considered in Section 5-4. Sometimes, atomic orbitals whose symmetries do not match and therefore remain unchanged in the molecule are also called nonbonding. For example, the s and d_{yz} orbitals of the preceding example are nonbonding with respect to each other. There are examples of both types of nonbonding orbitals later in this chapter.

In addition to symmetry, the second major factor that must be considered in forming molecular orbitals is the relative energy of the atomic orbitals. As shown in Figure 5-4, when the two atomic orbitals have the same energy, the resulting interaction is strong and the resulting molecular orbitals have energies well below (bonding) and above (antibonding) that of the original atomic orbitals. When the two atomic orbitals have quite different energies, the interaction is weak, and the resulting molecular orbitals have nearly the same energies and shapes as the original atomic orbitals. For example, although they have the same symmetry, $1s$ and $2s$ orbitals do not combine significantly in diatomic molecules such as N_2 because their energies are too far apart. As we will see, there is some interaction between $2s$ and $2p$, but it is relatively small. The general rule is that the closer the energy match, the stronger the interaction.

5-2 HOMONUCLEAR DIATOMIC MOLECULES

5-2-1 MOLECULAR ORBITALS

Although apparently satisfactory Lewis electron-dot diagrams of N_2, O_2, and F_2 can be drawn, the same is not true of Li_2, Be_2, B_2, and C_2, which cannot show the usual octet structure. In addition, the Lewis diagram of O_2 shows a simple double-bonded molecule, but experiment has shown it to have two unpaired electrons, making it paramagnetic (in fact, liquid oxygen poured between the poles of a large horseshoe magnet is attracted into the field and held there). As we will see, the molecular orbital description is more in agreement with experiment. Figure 5-5 shows the full set of molecular orbitals for the homonuclear diatomic molecules of the first 10 elements, with the energies appropriate for O_2. The diagram shows the order of energy levels for the molecular orbitals assuming interactions only between atomic orbitals of identical energy. The energies of the molecular orbitals change with increasing atomic number but the general pattern remains similar (with some subtle changes, as described in several examples that follow), even for heavier atoms lower in the periodic table. Electrons fill the molecular

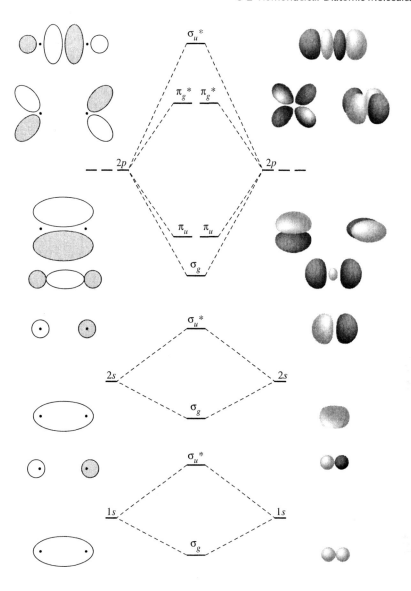

FIGURE 5-5 Molecular Orbitals for the First 10 Elements, with no σ-σ Interaction.

orbitals according to the same rules that govern the filling of atomic orbitals (filling from lowest to highest energy [aufbau], maximum spin multiplicity consistent with the lowest net energy [Hund's rules], and no two electrons with identical quantum numbers [Pauli exclusion principle]).

The overall number of bonding and antibonding electrons determines the number of bonds (bond order):

$$\text{Bond order} = \frac{1}{2}\left[\left(\begin{array}{c}\text{number of electrons}\\\text{in bonding orbitals}\end{array}\right) - \left(\begin{array}{c}\text{number of electrons}\\\text{in antibonding orbitals}\end{array}\right)\right]$$

For example, O_2, with 10 electrons in bonding orbitals and 6 electrons in antibonding orbitals, has a bond order of 2, a double bond. Counting only valence electrons (8 bonding and 4 antibonding) gives the same result. Because the molecular orbitals derived from the $1s$ orbitals have the same number of bonding and antibonding electrons, they have no net effect on the bonding.

Additional labels are helpful in describing the orbitals and have been added to Figure 5-5. We have added g and u subscripts, which are used as described at the end of

Section 4-3-3: g for *gerade*, orbitals symmetric to inversion, and u for *ungerade*, orbitals antisymmetric to inversion (those whose signs change on inversion). The g or u notation describes the symmetry of the orbitals without a judgment as to their relative energies.

EXAMPLE

Add a g or u label to each of the molecular orbitals in the energy level diagram in Figure 5-2. From top to bottom, the orbitals are σ_u^*, π_g^*, π_u, and σ_g.

EXERCISE 5-2

Add a g or u label to each of the molecular orbitals in Figure 5-3(a).

5-2-2 ORBITAL MIXING

So far, we have considered primarily interactions between orbitals of identical energy. However, orbitals with similar, but not equal, energies interact if they have appropriate symmetries. We will outline two approaches to analyzing this interaction, one in which the molecular orbitals interact and one in which the atomic orbitals interact directly.

When two molecular orbitals of the same symmetry have similar energies, they interact to lower the energy of the lower orbital and raise the energy of the higher. For example, in the homonuclear diatomics, the $\sigma_g(2s)$ and $\sigma_g(2p)$ orbitals both have σ_g symmetry (symmetric to infinite rotation and inversion); these orbitals interact to lower the energy of the $\sigma_g(2s)$ and to raise the energy of the $\sigma_g(2p)$, as shown in Figure 5-6(b).

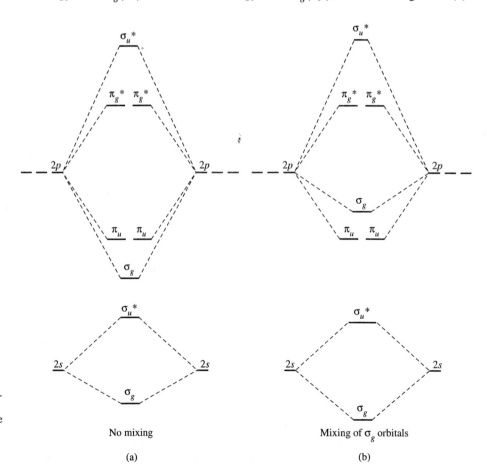

FIGURE 5-6 Interaction between Molecular Orbitals. Mixing molecular orbitals of the same symmetry results in a greater energy difference between the orbitals. The σ orbitals mix strongly; the σ^* orbitals differ more in energy and mix weakly.

No mixing

(a)

Mixing of σ_g orbitals

(b)

Similarly, the $\sigma_u^*(2s)$ and $\sigma_u^*(2p)$ orbitals interact to lower the energy of the $\sigma_u^*(2s)$ and to raise the energy of the $\sigma_u^*(2p)$. This phenomenon is called **mixing**. Mixing takes into account that molecular orbitals with similar energies interact if they have appropriate symmetry, a factor that has been ignored in Figure 5-5. When two molecular orbitals of the same symmetry mix, the one with higher energy moves still higher and the one with lower energy moves lower in energy.

Alternatively, we can consider that the four molecular orbitals (MOs) result from combining the four atomic orbitals (two $2s$ and two $2p_z$) that have similar energies. The resulting molecular orbitals have the following general form (where a and b identify the two atoms):

$$\Psi = c_1\psi(2s_a) \pm c_2\psi(2s_b) \pm c_3\psi(2p_a) \pm c_4\psi(2p_b)$$

For homonuclear molecules, $c_1 = c_2$ and $c_3 = c_4$ in each of the four MOs. The lowest energy MO has larger values of c_1 and c_2, the highest has larger values of c_3 and c_4, and the two intermediate MOs have intermediate values for all four coefficients. The symmetry of these four orbitals is the same as those without mixing, but their shapes are changed somewhat by having the mixture of s and p character. In addition, the energies are shifted, higher for the upper two and lower for the two lower energy orbitals.

As we will see, s-p mixing can have an important influence on the energy of molecular orbitals. For example, in the early part of the second period (Li_2 to N_2), the σ_g orbital formed from $2p$ orbitals is higher in energy than the π_u orbitals formed from the other $2p$ orbitals. This is an inverted order from that expected without mixing (Figure 5-6). For B_2 and C_2, this affects the magnetic properties of the molecules. In addition, mixing changes the bonding-antibonding nature of some of the orbitals. The orbitals with intermediate energies may have either slightly bonding or slightly anti-bonding character and contribute in minor ways to the bonding, but in some cases may be considered essentially nonbonding orbitals because of their small contribution and intermediate energy. Each orbital must be considered separately on the basis of the actual energies and electron distributions.

5-2-3 HOMONUCLEAR DIATOMIC MOLECULES

Before proceeding with examples of homonuclear diatomic molecules, it is necessary to define two types of magnetic behavior, **paramagnetic** and **diamagnetic**. Paramagnetic compounds are attracted by an external magnetic field. This attraction is a consequence of one or more unpaired electrons behaving as tiny magnets. Diamagnetic compounds, on the other hand, have no unpaired electrons and are repelled slightly by magnetic fields. (An experimental measure of the magnetism of compounds is the **magnetic moment**, a term that will be described further in Chapter 10 in the discussion of the magnetic properties of coordination compounds.)

H_2, He_2, and the homonuclear diatomic species shown in Figure 5-7 will be discussed in the following pages. In the progression across the periodic table, the energy of all the orbitals decreases as the increased nuclear charge attracts the electrons more strongly. As shown in Figure 5-7, the change is larger for σ orbitals than for π orbitals, resulting from the larger overlap of the atomic orbitals that participate in σ interactions. As shown in Figure 2-7, the atomic orbitals from which the σ orbitals are derived have higher electron densities near the nucleus.

H_2 $[\sigma_g^2(1s)]$

This the simplest of the diatomic molecules. The MO description (see Figure 5-1) shows a single σ bond containing one electron pair. The ionic species H_2^+, having a

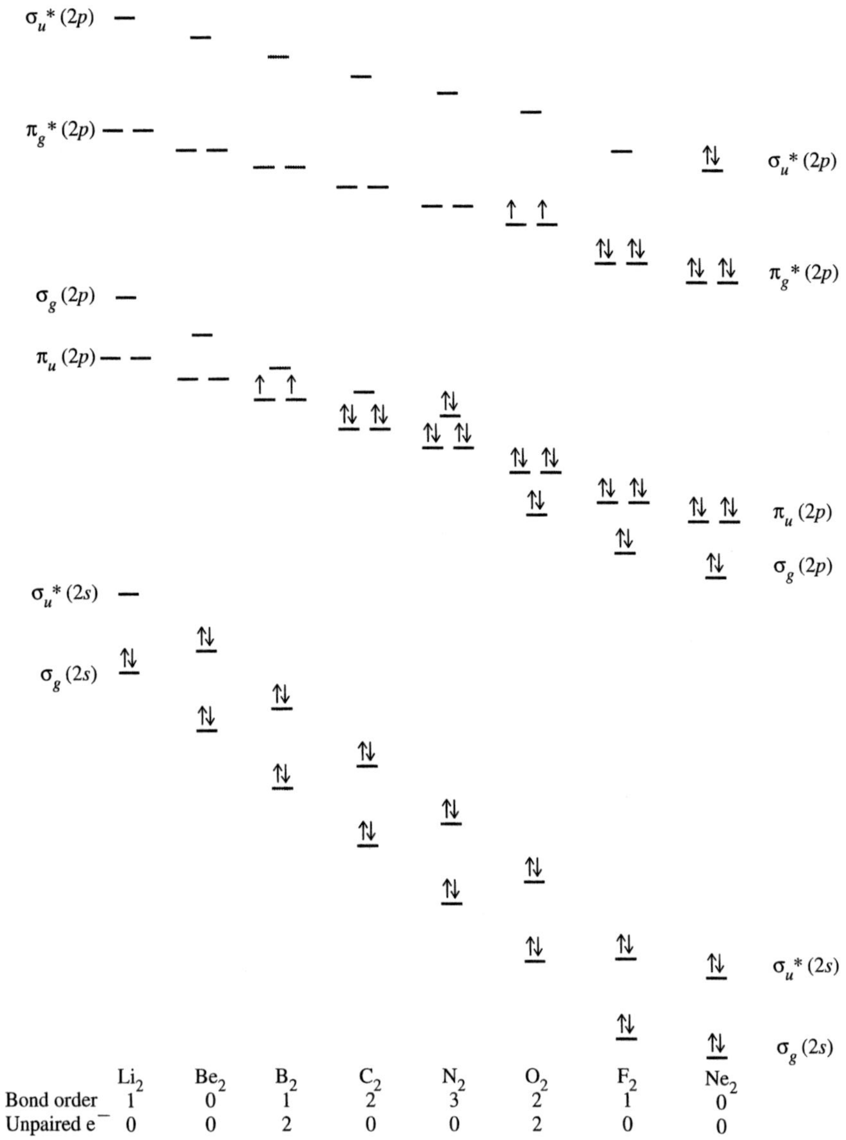

$\sigma_u^*(2p)$ —

$\pi_g^*(2p)$ — —

$\sigma_u^*(2p)$

$\sigma_g(2p)$ —

$\pi_u(2p)$ — —

$\pi_g^*(2p)$

$\pi_u(2p)$

$\sigma_g(2p)$

$\sigma_u^*(2s)$ —

$\sigma_g(2s)$

$\sigma_u^*(2s)$

$\sigma_g(2s)$

FIGURE 5-7 Energy Levels of the Homonuclear Diatomics of the Second Period.

	Li₂	Be₂	B₂	C₂	N₂	O₂	F₂	Ne₂
Bond order	1	0	1	2	3	2	1	0
Unpaired e⁻	0	0	2	0	0	2	0	0

bond order of $\frac{1}{2}$, has been detected in low-pressure gas discharge systems. As expected, it is less stable than H_2 and has a considerably longer bond distance (106 pm) than H_2 (74.2 pm).

He_2 [$\sigma_g{}^2\sigma_u{}^{*2}(1s)$]

The molecular orbital description of He_2 predicts two electrons in a bonding orbital and two electrons in an antibonding orbital, with a bond order of zero—in other words, no bond. This is what is observed experimentally. The noble gas He has no significant tendency to form diatomic molecules and, like the other noble gases, exists in the form of free atoms. He_2 has been detected only in very low pressure and low temperature molecular beams. It has a very low binding energy,[3] approximately 0.01 J/mol; for comparison, H_2 has a bond energy of 436 kJ/mol.

[3]F. Luo, G. C. McBane, G. Kim, C. F. Giese, and W. R. Gentry, *J. Chem. Phys.*, **1993**, *98*, 3564.

Li_2 $[\sigma_g^2(2s)]$

As shown in Figure 5-7, the MO model predicts a single Li—Li bond in Li_2, in agreement with gas phase observations of the molecule.

Be_2 $[\sigma_g^2\sigma_u^{*2}(2s)]$

Be_2 has the same number of antibonding and bonding electrons and consequently a bond order of zero. Hence, like He_2, Be_2 is not a stable chemical species.

B_2 $[\pi_u^1\pi_u^1(2p)]$

Here is an example in which the MO model has a distinct advantage over the Lewis dot picture. B_2 is found only in the gas phase; solid boron is found in several very hard forms with complex bonding, primarily involving B_{12} icosahedra. B_2 is paramagnetic. This behavior can be explained if its two highest energy electrons occupy separate π orbitals as shown. The Lewis dot model cannot account for the paramagnetic behavior of this molecule.

B_2 is also a good example of the energy level shift caused by the mixing of s and p orbitals. In the absence of mixing, the $\sigma_g(2p)$ orbital is expected to be lower in energy than the $\pi_u(2p)$ orbitals and the resulting molecule would be diamagnetic. However, mixing of the $\sigma_g(2s)$ orbital with the $\sigma_g(2p)$ orbital (see Figure 5-6) lowers the energy of the $\sigma_g(2s)$ orbital and increases the energy of the $\sigma_g(2p)$ orbital to a higher level than the π orbitals, giving the order of energies shown in Figure 5-7. As a result, the last two electrons are unpaired in the **degenerate** (having the same energy) π orbitals, and the molecule is paramagnetic. Overall, the bond order is 1, even though the two π electrons are in different orbitals.

C_2 $[\pi_u^2\pi_u^2(2p)]$

The simple MO picture of C_2 predicts a doubly bonded molecule with all electrons paired, but with both **highest occupied molecular orbitals (HOMOs)** having π symmetry. It is unusual because it has two π bonds and no σ bond. The bond dissociation energies of B_2, C_2, and N_2 increase steadily, indicating single, double, and triple bonds with increasing atomic number. Although C_2 is not a commonly encountered chemical species (carbon is more stable as diamond, graphite, and the fullerenes described in Chapter 8), the acetylide ion, C_2^{2-}, is well known, particularly in compounds with alkali metals, alkaline earths, and lanthanides. According to the molecular orbital model, C_2^{2-} should have a bond order of 3 (configuration $\pi_u^2\pi_u^2\sigma_g^2$). This is supported by the similar C—C distances in acetylene and calcium carbide (acetylide)[4,5]:

C—C *Distance (pm)*	
C=C (gas phase)	132
H—C≡C—H	120.5
CaC$_2$	119.1

[4]M. Atoji, *J. Chem. Phys.*, **1961**, *35*, 1950.

[5]J. Overend and H. W. Thompson, *Proc. R. Soc. London*, **1954**, *A234*, 306.

$N_2 [\pi_u^2 \pi_u^2 \sigma_g^2 (2p)]$

N_2 has a triple bond according to both the Lewis and the molecular orbital models. This is in agreement with its very short N—N distance (109.8 pm) and extremely high bond dissociation energy (942 kJ/mol). Atomic orbitals decrease in energy with increasing nuclear charge Z as shown in Figure 5-7; as the effective nuclear charge increases, all orbitals are pulled to lower energies. The shielding effect and electron-electron interactions described in Section 2-2-4 cause an increase in the difference between the $2s$ and $2p$ orbital energies as Z increases, from 5.7 eV for boron to 8.8 eV for carbon and 12.4 eV for nitrogen. (A table of these energies is given in Table 5-1 in Section 5-3-1.) As a result, the $\sigma_g(2s)$ and $\sigma_g(2p)$ levels of N_2 interact (mix) less than the B_2 and C_2 levels, and the $\sigma_g(2p)$ and $\pi_u(2p)$ are very close in energy. The order of energies of these orbitals has been a matter of controversy and will be discussed in more detail in Section 5-2-4 on photoelectron spectroscopy.[6]

$O_2 [\sigma_g^2 \pi_u^2 \pi_u^2 \pi_g^{*1} \pi_g^{*1} (2p)]$

O_2 is paramagnetic. This property, as for B_2, cannot be explained by the traditional Lewis dot structure ($\ddot{:}O{=}O\ddot{:}$), but is evident from the MO picture, which assigns two electrons to the degenerate π_g^* orbitals. The paramagnetism can be demonstrated by pouring liquid O_2 between the poles of a strong magnet; some of the O_2 will be held between the pole faces until it evaporates. Several ionic forms of diatomic oxygen are known, including O_2^+, O_2^-, and O_2^{2-}. The internuclear O—O distance can be conveniently correlated with the bond order predicted by the molecular orbital model, as shown in the following table.

	Bond Order	Internuclear Distance (pm)
O_2^+ (dioxygenyl)[7]	2.5	112.3
O_2 (dioxygen)[8]	2.0	120.07
O_2^- (superoxide)[9]	1.5	128
O_2^{2-} (peroxide)[8]	1.0	149

NOTE: Oxygen-oxygen distances in O_2^- and O_2^{2-} are influenced by the cation. This influence is especially strong in the case of O_2^{2-} and is one factor in its unusually long bond distance.

The extent of mixing is not sufficient in O_2 to push the $\sigma_g(2p)$ orbital to higher energy than the $\pi_g(2p)$ orbitals. The order of molecular orbitals shown is consistent with the photoelectron spectrum discussed in Section 5-2-4.

$F_2 [\sigma_g^2 \pi_u^2 \pi_u^2 \pi_g^{*2} \pi_g^{*2} (2p)]$

The MO picture of F_2 shows a diamagnetic molecule having a single fluorine-fluorine bond, in agreement with experimental data on this very reactive molecule.

The net bond order in N_2, O_2, and F_2 is the same whether or not mixing is taken into account, but the order of the filled orbitals is different. The switching of the order of the

---

[6]In the first and second editions of this text, the order of the σ_g and π_u orbitals in N_2 was reversed from the order in Figure 5-7. We have since become persuaded that the σ_g orbital has the higher energy.

[7]G. Herzberg, *Molecular Spectra and Molecular Structure I: The Spectra of Diatomic Molecules*, Van Nostrand-Reinhold, New York, 1950, p. 366.

[8]S. L. Miller and C. H. Townes, *Phys. Rev.*, **1953**, *90*, 537.

[9]N.-G. Vannerberg, *Prog. Inorg. Chem.*, **1963**, *4*, 125.

$\sigma_g(2p)$ and $\pi_u(2p)$ orbitals can occur because these orbitals are so close in energy; minor changes in either orbital can switch their order. The energy difference between the $2s$ and $2p$ orbitals of the atoms increases with increasing nuclear charge, from 5.7 eV in boron to 27.7 eV in fluorine (details are in Section 5-3-1). Because the difference becomes greater, the s-p interaction decreases and the "normal" order of molecular orbitals returns in O_2 and F_2. The higher σ_g orbital is seen again in CO, described later in Section 5-3-1.

Ne₂

All the molecular orbitals are filled, there are equal numbers of bonding and antibonding electrons, and the bond order is therefore zero. The Ne_2 molecule is a transient species, if it exists at all.

One triumph of molecular orbital theory is its prediction of two unpaired electrons for O_2. It had long been known that ordinary oxygen is paramagnetic, but the earlier bonding theories required use of a special "three-electron bond"[10] to explain this phenomenon. On the other hand, the molecular orbital description provides for the unpaired electrons directly. In the other cases described previously, the experimental facts (paramagnetic B_2, diamagnetic C_2) require a shift of orbital energies, raising σ_g above π_u, but they do not require addition of any different type of orbitals or bonding. Once the order has been determined experimentally, molecular calculations can be tested against the experimental results to complete the picture.

Bond lengths in homonuclear diatomic molecules

Figure 5-8 shows the variation of bond distance with the number of valence electrons in second-period p block homonuclear diatomic molecules. As the number of electrons increases, the number in bonding orbitals also increases, the bond strength becomes greater, and the bond length becomes shorter. This continues up to 10 valence electrons in N_2 and then the trend reverses because the additional electrons occupy antibonding orbitals. The ions N_2^+, O_2^+, O_2^-, and O_2^{2-} are also shown in the figure and follow a similar trend.

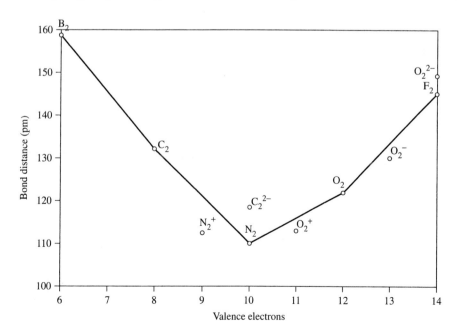

FIGURE 5-8 Bond Distances of Homonuclear Diatomic Molecules and Ions.

[10]L. Pauling, *The Nature of the Chemical Bond*, 3rd ed., Cornell University Press, Ithaca, NY, 1960, pp. 340–354.

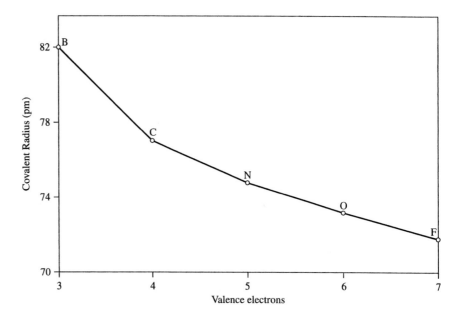

FIGURE 5-9 Covalent Radii of Second-Period Atoms.

Figure 5-9 shows the change in covalent radius for the atoms, decreasing as the number of valence electrons increases because the increasing nuclear charge pulls the electrons closer to the nucleus. For the elements boron through nitrogen, the trends shown in Figures 5-8 and 5-9 are similar: as the covalent radius of the atom decreases, the bond distance of the matching diatomic molecule also decreases. However, beyond nitrogen these trends diverge. Even though the covalent radii continue to decrease (N > O > F), the bond distances in their diatomic molecules increase ($N_2 < O_2 < F_2$) because of the increasing population of antibonding orbitals.

Similarly, if there were no other influences, the H—F bond should be shorter than the H—O bond, and so on across the period. In fact, the bond distances are H—B, 120 pm; H—C, 109 pm; H—N, 101.2 pm; H—O, 96 pm; and H—F, 91.8 pm. The trend is consistent with the curve in Figure 5-9, although the changes in bond distance between species are larger than expected from the covalent radii. Figure 5-8 shows still larger differences because of the additional bonds; double and triple bonds have much shorter bond distances than single bonds, regardless of the average covalent radius of the component atoms.

5-2-4 PHOTOELECTRON SPECTROSCOPY

In addition to data on bond distances and energies, specific information about the energies of electrons in orbitals can be determined from photoelectron spectroscopy,[11] one of the more direct methods for determining orbital energies. In this technique, ultraviolet (UV) light or X-rays dislodge electrons from molecules:

$$O_2 + h\nu \text{ (photons)} \longrightarrow O_2{}^+ + e^-$$

The kinetic energy of the expelled electrons can be measured; the difference between the energy of the incident photons and this kinetic energy equals the ionization energy (binding energy) of the electron:

$$\text{Ionization energy} = h\nu(\text{photons}) - \text{kinetic energy of expelled electron}$$

[11]E. A. V. Ebsworth, D. W. H. Rankin, and S. Cradock, *Structural Methods in Inorganic Chemistry*, 2nd ed., CRC Press, Boca Raton, FL, pp. 255–279. Pages 274 and 275 discuss the spectra of N_2 and O_2.

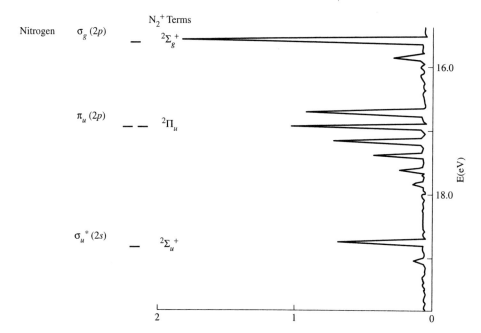

FIGURE 5-10 Photoelectron Spectrum and Molecular Orbital Energy Levels of N_2. (Photoelectron spectrum reproduced with permission from J. L. Gardner and J. A. R. Samson, *J. Chem. Phys.*, **1975**, *62*, 1447.)

Ultraviolet light removes outer electrons, usually from gases; X-rays are more energetic and remove inner electrons as well, from any physical state. Figures 5-10 and 5-11 show photoelectron spectra for N_2 and O_2 and the relative energies of the highest occupied orbitals of the ions. The lower energy peaks (at the top in the figure) are for the higher energy orbitals (less energy required to remove electrons). If the energy levels of the ionized molecule are assumed to be essentially the same as those of the uncharged molecule, the observed energies can be directly correlated with the molecular orbital energies. The levels in the N_2 spectrum are much closer together than in the O_2

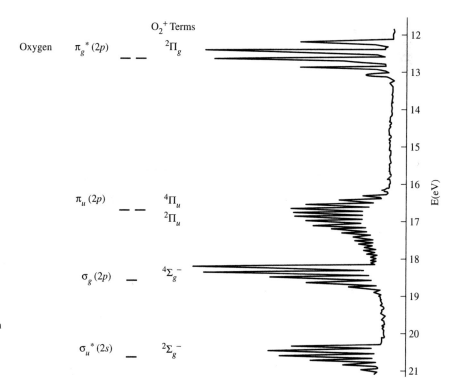

FIGURE 5-11 Photoelectron Spectrum and Molecular Orbital Energy Levels of O_2. (Photoelectron spectrum reproduced with permission from J. H. D. Eland, *Photoelectron Spectroscopy*, Butterworths, London, 1974, p. 10.)

spectrum (about 3 eV between the first and third major peaks in N_2, about 6 eV for the corresponding difference in O_2), and some theoretical calculations have disagreed about the order of the highest occupied orbitals. A recent paper[12] compared different calculation methods and showed that the different order of energy levels was simply a consequence of the method of calculation used; the methods favored by the authors agree with the experimental results, with σ_g above π_u.

The photoelectron spectrum shows the π_u lower (Figure 5-10). In addition to the ionization energies of the orbitals, the spectrum shows the interaction of the electronic energy with the vibrational energy of the molecule. Because vibrational energy levels are much closer in energy than electronic levels, any collection of molecules has an energy distribution through many different vibrational levels. Because of this, transitions between electronic levels also include transitions between different vibrational levels, resulting in multiple peaks for a single electronic transition. Orbitals that are strongly involved in bonding have vibrational fine structure (multiple peaks); orbitals that are less involved in bonding have only a few individual peaks at each energy level.[13] The N_2 spectrum indicates that the π_u orbitals are more involved in the bonding than either of the σ orbitals. The CO photoelectron spectrum (Figure 5-14) has a similar pattern. The O_2 photoelectron spectrum (Figure 5-11) has much more vibrational fine structure for all the energy levels, with the π_u levels again more involved in bonding than the other orbitals.

The photoelectron spectra of O_2 (Figure 5-11) and of CO (Figure 5-14) show the expected order of energy levels. The vibrational fine structure indicates that all the orbitals are important to bonding in the molecules.

5-2-5 CORRELATION DIAGRAMS

Mixing of orbitals of the same symmetry, as in the examples of Section 5-2-3, is seen in many other molecules. A **correlation diagram**[14] for this phenomenon is shown in Figure 5-12. This diagram shows the calculated effect of moving two atoms together, from a large interatomic distance on the right, with no interatomic interaction, to zero interatomic distance on the left, where the two nuclei become, in effect, a single nucleus. The simplest example has two hydrogen atoms on the right and a helium atom on the left. Naturally, such merging of two atoms into one never happens outside the realm of high-energy physics, but we consider the orbital changes as if it could. The diagram shows how the energies of the orbitals change with the internuclear distance and change from the order of atomic orbitals on the left to the order of molecular orbitals of similar symmetry on the right.

On the right are the usual atomic orbitals—$1s$, $2s$, and $2p$ for each of the two separated atoms. As the atoms approach each other, their atomic orbitals interact to form molecular orbitals.[15] The $1s$ orbitals form $1\sigma_g$ and $1\sigma_u^*$, $2s$ form $2\sigma_g$ and $2\sigma_u^*$, and $2p$ form $3\sigma_g$, $1\pi_u$, $1\pi_g^*$, and $3\sigma_u^*$. As the atoms move closer together (toward the left in the diagram), the bonding MOs decrease in energy, while the antibonding MOs increase in energy. At the far left, the MOs become the atomic orbitals of a united atom with twice the nuclear charge.

[12]R. Stowasser and R. Hoffmann, *J. Am. Chem. Soc.*, **1999**, *121*, 3414.

[13]R. S. Drago, *Physical Methods in Chemistry*, 2nd ed., Saunders College Publishing, Philadelphia, 1992, pp. 671–677.

[14]R. McWeeny, *Coulson's Valence*, 3rd Ed., Oxford University Press, Oxford, 1979, pp. 97–103.

[15]Molecular orbitals are labeled in many different ways. Most in this book are numbered within each set of the same symmetry ($1\sigma_g$, $2\sigma_g$ and $1\sigma_u^*$, $2\sigma_u^*$). In some figures, $1\sigma_g$ and $1\sigma_u^*$ MOs from $1s$ atomic orbitals are understood to be at lower energies than the MOs shown and are omitted.

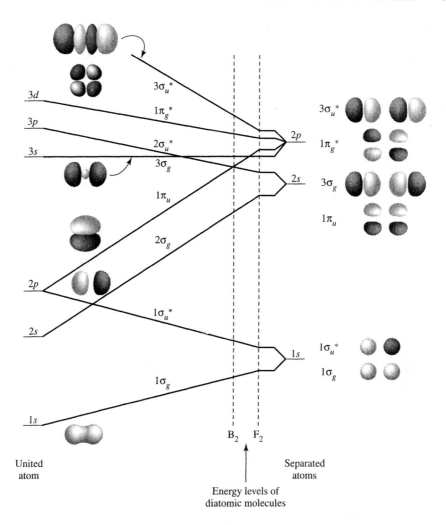

FIGURE 5-12 Correlation Diagram for Homonuclear Diatomic Molecular Orbitals.

Symmetry is used to connect the molecular orbitals with the atomic orbitals of the united atom. Consider the $1\sigma_u{}^*$ orbital as an example. It is formed as the antibonding orbital from two $1s$ orbitals, as shown on the right side of the diagram. It has the same symmetry as a $2p_z$ atomic orbital (where z is the axis through both nuclei), which is the limit on the left side of the diagram. The degenerate $1\pi_u$ MOs are also connected to the $2p$ orbitals of the united atom, because they have the same symmetry as a $2p_x$ or $2p_y$ orbital (see Figure 5-2).

As another example, the degenerate pair of $1\pi_g{}^*$ MOs, formed by the difference of the $2p_x$ or $2p_y$ orbitals of the separate atoms, is connected to the $3d$ orbitals on the left side because the $1\pi_g{}^*$ orbitals have the same symmetry as the d_{xz} or d_{yz} orbitals (see Figure 5-2). The π orbitals formed from p_x and p_y orbitals are degenerate (have the same energy), as are the p orbitals of the merged atom, and the π^* orbitals from the same atomic orbitals are degenerate, as are the d orbitals of the merged atom.

Another consequence of this phenomenon is called the noncrossing rule, which states that orbitals of the same symmetry interact so that their energies never cross.[16] This rule helps in assigning correlations. If two sets of orbitals of the same symmetry seem to result in crossing in the correlation diagram, the matchups must be changed to prevent it.

[16]C. J. Ballhausen and H. B. Gray, *Molecular Orbital Theory*, W. A. Benjamin, New York, 1965, pp. 36–38.

The actual energies of molecular orbitals for diatomic molecules are intermediate between the extremes of this diagram, approximately in the region set off by the vertical lines. Toward the right within this region, closer to the separated atoms, the energy sequence is the "normal" one of O_2 and F_2; further to the left, the order of molecular orbitals is that of B_2, C_2 and N_2, with $\sigma_g(2p)$ above $\pi_u(2p)$.

5-3 HETERONUCLEAR DIATOMIC MOLECULES

5-3-1 POLAR BONDS

Heteronuclear diatomic molecules follow the same general bonding pattern as the homonuclear molecules described previously, but a greater nuclear charge on one of the atoms lowers its atomic energy levels and shifts the resulting molecular orbital levels. In dealing with heteronuclear molecules, it is necessary to have a way to estimate the energies of the atomic orbitals that may interact. For this purpose, the orbital potential energies, given in Table 5-1 and Figure 5-13, are useful. These potential energies are negative, because they represent attraction between valence electrons and atomic nuclei. The values are the average energies for all electrons in the same level (for example, all $3p$ electrons), and are weighted averages of all the energy states possible. These

TABLE 5-1
Orbital Potential Energies

| Atomic Number | Element | Orbital Potential Energy (eV) | | | | | | |
		1s	2s	2p	3s	3p	4s	4p
1	H	−13.61						
2	He	−24.59						
3	Li		−5.39					
4	Be		−9.32					
5	B		−14.05	−8.30				
6	C		−19.43	−10.66				
7	N		−25.56	−13.18				
8	O		−32.38	−15.85				
9	F		−40.17	−18.65				
10	Ne		−48.47	−21.59				
11	Na				−5.14			
12	Mg				−7.65			
13	Al				−11.32	−5.98		
14	Si				−15.89	−7.78		
15	P				−18.84	−9.65		
16	S				−22.71	−11.62		
17	Cl				−25.23	−13.67		
18	Ar				−29.24	−15.82		
19	K						−4.34	
20	Ca						−6.11	
30	Zn						−9.39	
31	Ga						−12.61	−5.93
32	Ge						−16.05	−7.54
33	As						−18.94	−9.17
34	Se						−21.37	−10.82
35	Br						−24.37	−12.49
36	Kr						−27.51	−14.22

SOURCE: J. B. Mann, T. L. Meek, and L. C. Allen, *J. Am. Chem. Soc.*, **2000**, *122*, 2780.

NOTE: All energies are negative, representing average attractive potentials between the electrons and the nucleus for all terms of the specified orbitals.

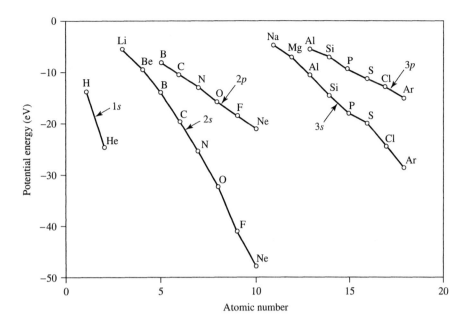

FIGURE 5-13 Orbital Potential Energies.

states are called terms and are explained in Chapter 11. For this reason, the values do not show the variations of the ionization energies seen in Figure 2-10, but steadily become more negative from left to right within a period, as the increasing nuclear charge attracts all the electrons more strongly.

The atomic orbitals of homonuclear diatomic molecules have identical energies, and both atoms contribute equally to a given MO. Therefore, in the equations for the molecular orbitals, the coefficients for the two atomic orbitals are identical. In heteronuclear diatomic molecules such as CO and HF, the atomic orbitals have different energies and a given MO receives unequal contributions from the atomic orbitals; the equation for that MO has a different coefficient for each of the atomic orbitals that compose it. As the energies of the atomic orbitals get farther apart, the magnitude of the interaction decreases. The atomic orbital closer in energy to an MO contributes more to the MO, and its coefficient is larger in the wave equation.

The molecular orbitals of CO are shown in Figure 5-14. CO has $C_{\infty v}$ symmetry, but the p_x and p_y orbitals have C_{2v} symmetry if the signs of the orbital lobes are ignored as in the diagram (the signs are ignored only for the purpose of choosing a point group, but must be included for the rest of the process). Using the C_{2v} point group rather than $C_{\infty v}$ simplifies the orbital analysis by avoiding the infinite rotation axis of $C_{\infty v}$. The s and p_z group orbitals have A_1 symmetry and form molecular orbitals with σ symmetry; the p_x and p_y group orbitals have B_1 and B_2 symmetry, respectively (the p_x and p_y orbitals change sign on C_2 rotation and change sign on one σ_v reflection, but not on the other), and form π orbitals. When combined to form molecular orbitals, the B_1 and B_2 orbitals have the same energy, behaving like the E_1 representation of the $C_{\infty v}$ group.

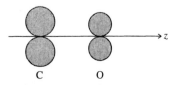

Diagram of C_{2v} symmetry of p orbitals

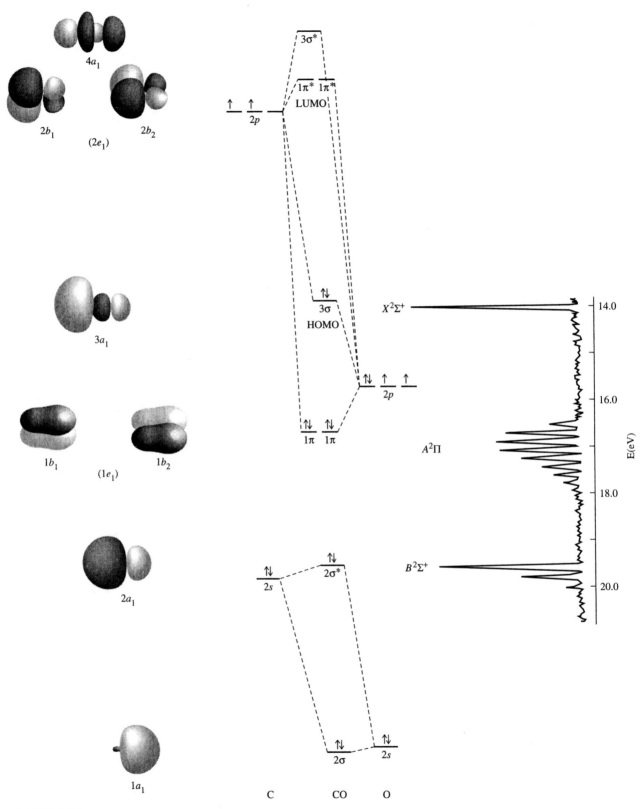

FIGURE 5-14 Molecular Orbitals and Photoelectron Spectrum of CO. Molecular orbitals 1σ and $1\sigma^*$ are from the $1s$ orbitals and are not shown, The e_1 and e_2 labels in the left-hand column are for the $C_{\infty v}$ symmetry labels; the b_1 and b_2 labels are for C_2v symmetry. (Photoelectron spectrum reproduced with permission from J. L. Gardner and J. A. R. Samson, *J. Chem. Phys.*, **1975**, *62*, 1447.)

The bonding orbital 2σ has more contribution from (and is closer in energy to) the lower energy oxygen $2s$ atomic orbital; the antibonding $2\sigma^*$ orbital has more contribution from (and is closer in energy to) the higher energy carbon $2s$ atomic orbital. In the simplest case, the bonding orbital is nearly the same in energy and shape as the lower energy atomic orbital, and the antibonding orbital is nearly the same in energy and shape as the higher energy atomic orbital. In more complicated cases (such as the $2\sigma^*$ orbital of CO) other orbitals (the oxygen $2p_z$ orbital) contribute, and the orbital shapes and energies are not as easily predicted. As a practical matter, atomic orbitals with energy differences greater than 12 or 13 eV usually do not interact significantly.

Mixing of the two σ levels and the two σ^* levels, like that seen in the homonuclear σ_g and σ_u orbitals, causes a larger split in energy between them, and the 3σ is higher than the π levels. The p_x and p_y orbitals also form four molecular π orbitals, two bonding and two antibonding. When the electrons are filled in as in Figure 5-14, the valence orbitals form four bonding pairs and one antibonding pair for a net of three bonds.

EXAMPLE

Molecular orbitals for HF can be found by using the techniques just described. The symmetry of the molecule is $C_{\infty v}$, which can be simplified to C_{2v}, just as in the CO case. The $2s$ orbital of the F atom has an energy about 27 eV lower than that of the hydrogen $1s$, so there is very little interaction between them. The F orbital retains a pair of electrons. The F $2p_z$ orbital and the H $1s$, on the other hand, have similar energies and matching A_1 symmetries, allowing them to combine into bonding σ and antibonding σ^* orbitals. The F $2p_x$ and $2p_y$ orbitals have B_1 and B_2 symmetries and remain nonbonding, each with a pair of electrons. Overall, there is one bonding pair of electrons and three lone pairs.

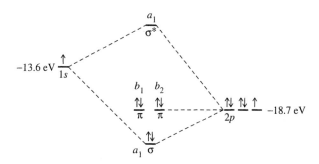

EXERCISE 5-3

Use similar arguments to explain the bonding in the OH$^-$ ion.

The molecular orbitals that will be of greatest interest for reactions between molecules are the **highest occupied molecular orbital (HOMO)** and the **lowest unoccupied molecular orbital (LUMO)**, collectively known as **frontier orbitals** because they lie at the occupied-unoccupied frontier. The MO diagram of CO helps explain its reaction chemistry with transition metals, which is not that predicted by simple electronegativity arguments that place more electron density on the oxygen. If this were true, metal carbonyls should bond as M—O—C, with the negative oxygen attached to the positive metal. The actual bonding is in the order M—C—O. The HOMO of CO is 3σ, with a higher electron density and a larger lobe on the carbon. The lone pair in this orbital forms a bond with a vacant orbital on the metal. The interaction between CO and metal orbitals is enormously important in the field of organometallic chemistry and will be discussed in detail in Chapter 13.

In simple cases, bonding MOs have a greater contribution from the lower energy atomic orbital, and their electron density is concentrated on the atom with the lower energy levels or higher electronegativity (see Figure 5-14). If this is so, why does the HOMO of CO, a bonding MO, have greater electron density on carbon, which has the higher energy levels? The answer lies in the way the atomic orbital contributions are divided. The p_z of oxygen has an energy that enables it to contribute to the $2\sigma^*$, the 3σ (the HOMO), and the $3\sigma^*$ MOs. The higher energy carbon p_z, however, only contributes significantly to the latter two. Because the p_z of the oxygen atom is divided among three MOs, it has a relatively weaker contribution to each one, and the p_z of the carbon atom has a relatively stronger contribution to each of the two orbitals to which it contributes.

The LUMOs are the $2\pi^*$ orbitals and are concentrated on carbon, as expected. The frontier orbitals can contribute electrons (HOMO) or accept electrons (LUMO) in reactions. Both are important in metal carbonyl bonding, which will be discussed in Chapter 13.

5-3-2 IONIC COMPOUNDS AND MOLECULAR ORBITALS

Ionic compounds can be considered the limiting form of polarity in heteronuclear diatomic molecules. As the atoms differ more in electronegativity, the difference in energy of the orbitals also increases, and the concentration of electrons shifts toward the more electronegative atom. At this limit, the electron is transferred completely to the more electronegative atom to form a negative ion, leaving a positive ion with a high-energy vacant orbital. When two elements with a large difference in their electronegativities (such as Li and F) combine, the result is an ionic compound. However, in molecular orbital terms, we can also consider an ion pair as if it were a covalent compound. In Figure 5-15, the atomic orbitals and an approximate indication of molecular orbitals for such a diatomic molecule are given. On formation of the compound LiF, the electron from the Li $2s$ orbital is transferred to the F $2p$ orbital, and the energy level of the $2p$ orbital is lowered.

In a more accurate picture of ionic crystals, the ions are held together in a three-dimensional lattice by a combination of electrostatic attraction and covalent bonding. Although there is a small amount of covalent character in even the most ionic compounds, there are no directional bonds, and each Li^+ ion is surrounded by six F^- ions, each of which in turn is surrounded by six Li^+ ions. The crystal molecular orbitals form energy bands, described in Chapter 7.

Formation of the ions can be described as a sequence of elementary steps, beginning with solid Li and gaseous F_2:

$Li(s) \longrightarrow Li(g)$	161 kJ/mol	(sublimation)
$Li(g) \longrightarrow Li^+(g) + e^-$	531 kJ/mol	(ionization, IE)
$\frac{1}{2}F_2(g) \longrightarrow F(g)$	79 kJ/mol	(dissociation)
$F(g) + e^- \longrightarrow F^-(g)$	-328 kJ/mol	(ionization, $-$EA)
$Li(s) + \frac{1}{2}F_2(g) \longrightarrow Li^+(g) + F^-(g)$	443 kJ/mole	

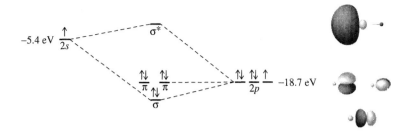

FIGURE 5-15 Approximate LiF Molecular Orbitals.

In order for a reaction to proceed spontaneously, the free energy change $(\Delta G = \Delta H - T\Delta S)$ must be negative. Although the entropy change for this reaction is positive, the very large positive ΔH results in a positive ΔG. If this were the final result, Li^+ and F^- would not react. However, the large attraction between the ions results in the release of 709 kJ/mol on formation of a single Li^+F^- ion pair, and 1239 kJ/mol on formation of a crystal:

$$Li^+(g) + F^-(g) \longrightarrow LiF(g) \qquad\qquad -709 \text{ kJ/mole (ion pairs)}$$

$$Li^+(g) + F^-(g) \longrightarrow LiF(s) \qquad\qquad -1239 \text{ kJ/mole (lattice enthalpy)}$$

The **lattice enthalpy** for crystal formation is large enough to overcome all the endothermic processes (and the negative entropy change) and to make formation of LiF from the elements a very favorable reaction.

5-4

MOLECULAR ORBITALS FOR LARGER MOLECULES

The methods described previously for diatomic molecules can be extended to obtain molecular orbitals for molecules consisting of three or more atoms, but more complex cases benefit from the use of formal methods of group theory. The process uses the following steps:

1. Determine the point group of the molecule. If it is a linear molecule, substituting a simpler point group that retains the symmetry of the orbitals (ignoring the signs) makes the process easier. Substitute D_{2h} for $D_{\infty h}$, C_{2v} for $C_{\infty v}$. This substitution retains the symmetry of the orbitals without the infinite-fold rotation axis.

2. Assign x, y, and z coordinates to the atoms, chosen for convenience. Experience is the best guide here. **The general rule in all the examples in this book is that the highest order rotation axis of the molecule is chosen as the z axis of the central atom.** In nonlinear molecules, the y axes of the outer atoms are chosen to point toward the central atom.

3. Find the characters of the representation for the combination of the $2s$ orbitals on the outer atoms and then repeat the process, finding the representations for each of the other sets of orbitals (p_x, p_y, and p_z). Later, these will be combined with

the appropriate orbitals of the central atom. As in the case of the vectors described in Chapter 4, any orbital that changes position during a symmetry operation contributes zero to the character of the resulting representation, any orbital that remains in its original position contributes 1, and any orbital that remains in the original position with the signs of its lobes reversed contributes -1.

4. Reduce each representation from Step 3 to its irreducible representations. This is equivalent to finding the symmetry of the **group orbitals** or the **symmetry-adapted linear combinations (SALCs)** of the orbitals. The group orbitals are then the combinations of atomic orbitals that match the symmetry of the irreducible representations.

5. Find the atomic orbitals of the central atom with the same symmetries (irreducible representations) as those found in Step 4.

6. Combine the atomic orbitals of the central atom and those of the group orbitals with the same symmetry and similar energy to form molecular orbitals. The total number of molecular orbitals formed equals the number of atomic orbitals used from all the atoms.[17]

In summary, the process used in creating molecular orbitals is to match the symmetries of the group orbitals (using their irreducible representations) with the symmetries of the central atom orbitals. If the symmetries match and the energies are not too different, there is an interaction (both bonding and antibonding); if not, there is no interaction.

The process can be carried further to obtain numerical values of the coefficients of the atomic orbitals used in the molecular orbitals.[18] For the qualitative pictures we will describe, it is sufficient to say that a given orbital is primarily composed of one of the atomic orbitals or that it is composed of roughly equal contributions from each of several atomic orbitals. The coefficients may be small or large, positive or negative, similar or quite different, depending on the characteristics of the orbital under consideration. Several computer software packages are available that will calculate these coefficients and generate the pictorial diagrams that describe the molecular orbitals.

5-4-1 FHF⁻

FHF^-, an example of very strong hydrogen bonding,[19] is a linear ion. FHF^- has $D_{\infty h}$ symmetry, but the infinite rotation axis of the $D_{\infty h}$ point group is difficult to work with. In cases like this, it is possible to use a simpler point group that still retains the symmetry of the orbitals. D_{2h} works well in this case, so it will be used for the rest of this section (see Section 5-3-1 for a similar choice for CO). The character table of this group shows the symmetry of the orbitals as well as the coordinate axes. For example, B_{1u} has the symmetry of the z axis and of the p_z orbitals on the fluorines; they are unchanged by the E, $C_2(z)$, $\sigma(xz)$, and $\sigma(yz)$ operations, and the $C_2(y)$, $C_2(x)$, i, and $\sigma(xy)$ operations change their signs.

---

[17]We use lower case labels on the molecular orbitals, with upper case for the atomic orbitals and for representations in general. This practice is common, but not universal.

[18]F. A. Cotton, *Chemical Applications of Group Theory*, 3rd ed., John Wiley & Sons, New York, 1990, pp. 133–188.

[19]J. H. Clark, J. Emsley, D. J. Jones, and R. E. Overill, *J. Chem. Soc.*, **1981**, 1219; J. Emsley, N. M. Reza, H. M. Dawes, and M. B. Hursthouse, *J. Chem. Soc. Dalton Trans.*, **1986**, 313.

D_{2h}	E	$C_2(z)$	$C_2(y)$	$C_2(x)$	i	$\sigma(xy)$	$\sigma(xz)$	$\sigma(yz)$		
A_g	1	1	1	1	1	1	1	1		x^2, y^2, z^2
B_{1g}	1	1	-1	-1	1	1	-1	-1	R_z	xy
B_{2g}	1	-1	1	-1	1	-1	1	-1	R_y	xz
B_{3g}	1	-1	-1	1	1	-1	-1	1	R_x	yz
A_u	1	1	1	1	-1	-1	-1	-1		
B_{1u}	1	1	-1	-1	-1	-1	1	1	z	
B_{2u}	1	-1	1	-1	-1	1	-1	1	y	
B_{3u}	1	-1	-1	1	-1	1	1	-1	x	

The axes used and the fluorine atom group orbitals are given in Figure 5-16; they are the $2s$ and $2p$ orbitals of the fluorine atoms, considered as pairs. These are the same combinations that formed bonding and antibonding orbitals in diatomic molecules (e.g., $p_{xa} + p_{xb}, p_{xa} - p_{xb}$), but they are now separated by the central H atom. As usual, we need to consider only the valence atomic orbitals ($2s$ and $2p$). The orbitals are numbered 1 through 8 for easier reference. The symmetry of each group orbital (SALC) can be found by comparing its behavior with each symmetry operation with the irreducible representations of the character table. The symmetry labels in Figure 5-16 show the results. For example, the $2s$ orbitals on the fluorine atoms give the two group orbitals 1 and 2. The designation "group orbital" does not imply direct bonding between the two fluorine atoms. Instead, group orbitals should be viewed merely as sets of similar orbitals. As before, the number of orbitals is always conserved, so the number of group orbitals is the same as the number of atomic orbitals combined to form them. We will now consider how these group orbitals may interact with atomic orbitals on the central atom, with each group orbital being treated in the same manner as an atomic orbital.

Atomic orbitals and group orbitals of the same symmetry can combine to form molecular orbitals, just as atomic orbitals of the same symmetry can combine to form

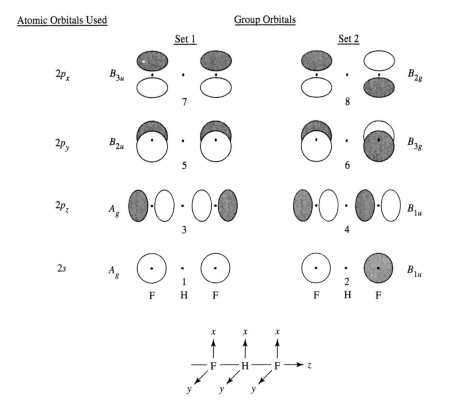

FIGURE 5-16 Group Orbitals for FHF$^-$.

group orbitals. Interaction of the A_g 1s orbital of hydrogen with the A_g orbitals of the fluorine atoms (group orbitals 1 and 3) forms bonding and antibonding orbitals, as shown in Figure 5-17. The overall set of molecular orbitals is shown in Figure 5-18.

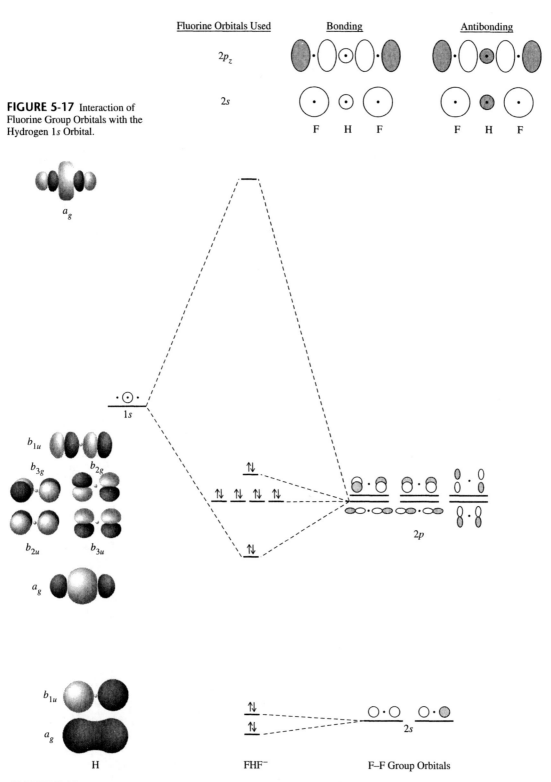

FIGURE 5-17 Interaction of Fluorine Group Orbitals with the Hydrogen 1s Orbital.

FIGURE 5-18 Molecular Orbital Diagram of FHF⁻.

Both sets of interactions are permitted by the symmetry of the orbitals involved. However, the energy match of the $1s$ orbital of hydrogen (orbital potential energy $= -13.6$ eV) is much better with the $2p_z$ of fluorine (-18.7 eV) than with the $2s$ of fluorine (-40.2 eV). Consequently, the $1s$ orbital of hydrogen interacts much more strongly with group orbital 3 than with group orbital 1 (Figure 5-18). Although the orbital potential energies of the H $1s$ and F $2s$ orbitals differ by almost 30 eV, some calculations show a small interaction between them.

In sketching the molecular orbital energy diagrams of polyatomic species, we will show the orbitals of the central atom on the far left, the group orbitals of the surrounding atoms on the far right, and the resulting molecular orbitals in the middle.

Five of the six group orbitals derived from the $2p$ orbitals of the fluorines do not interact with the central atom; these orbitals remain essentially nonbonding and contain lone pairs of electrons. There is a slight interaction between orbitals on non-neighboring atoms, but not enough to change their energies significantly. The sixth $2p$ group orbital, the $2p_z$ group orbital (number 3), interacts with the $1s$ orbital of hydrogen to give two molecular orbitals, one bonding and one antibonding. An electron pair occupies the bonding orbital. The group orbitals from the $2s$ orbitals of the fluorine atoms are much lower in energy than the $1s$ orbital of the hydrogen atom and are essentially nonbonding.

The Lewis approach to bonding requires two electrons to represent a single bond between two atoms and would result in four electrons around the hydrogen atom of FHF^-. The molecular orbital picture is more successful, with a 2-electron bond delocalized over *three* atoms (a 3-center, 2-electron bond). The bonding MO in Figures 5-17 and 5-18 shows how the molecular orbital approach represents such a bond: two electrons occupy a low-energy orbital formed by the interaction of all three atoms (a central atom and a two-atom group orbital). The remaining electrons are in the group orbitals derived from the p_x and p_y orbitals of the fluorine, at essentially the same energy as that of the atomic orbitals.

In general, larger molecular orbitals (extending over more atoms) have lower energies. Bonding molecular orbitals derived from three or more atoms, like the one in Figure 5-18, usually have lower energies than those that include molecular orbitals from only two atoms, but the total energy of a molecule is the sum of the energies of all of the electrons in all the orbitals. FHF^- has a bond energy of 212 kJ/mol and F—H distances of 114.5 pm. HF has a bond energy of 574 kJ/mol and an F—H bond distance of 91.7 pm.[20]

EXERCISE 5-4

Sketch the energy levels and the molecular orbitals for the H_3^+ ion, using linear geometry. Include the symmetry labels for the orbitals.

5-4-2 CO_2

Carbon dioxide, another linear molecule, has a more complicated molecular orbital description than FHF^-. Although the group orbitals for the oxygen atoms are identical to the group orbitals for the fluorine atoms in FHF^-, the central carbon atom in CO_2 has both s and p orbitals capable of interacting with the $2p$ group orbitals on the oxygen atoms. As in the case of FHF^-, CO_2 has $D_{\infty h}$ symmetry, but the simpler D_{2h} point group will be used.

[20]M. Mautner, *J. Am. Chem. Soc.*, **1984**, *106*, 1257.

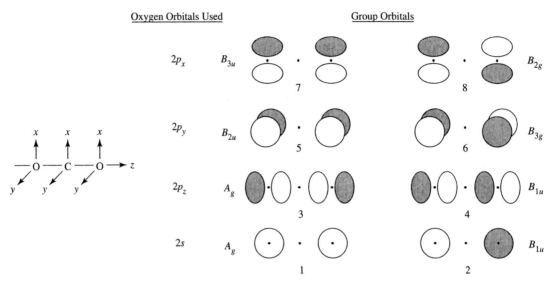

FIGURE 5-19 Group Orbital Symmetry in CO_2.

The group orbitals of the oxygen atoms are the same as those for the fluorine atoms shown in Figure 5-16. To determine which atomic orbitals of carbon are of correct symmetry to interact with the group orbitals, we will consider each of the group orbitals in turn. The combinations are shown again in Figure 5-19 and the carbon atomic orbitals are shown in Figure 5-20 with their symmetry labels for the D_{2h} point group.

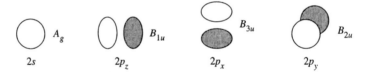

FIGURE 5-20 Symmetry of the Carbon Atomic Orbitals in the D_{2h} Point Group.

Group orbitals 1 and 2 in Figure 5-21, formed by adding and subtracting the oxygen $2s$ orbitals, have A_g and B_{1u} symmetry, respectively. Group orbital 1 is of appropriate symmetry to interact with the $2s$ orbital of carbon (both have A_g symmetry), and group orbital 2 is of appropriate symmetry to interact with the $2p_z$ orbital of carbon (both have B_{1u} symmetry).

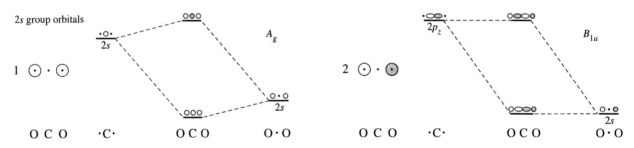

FIGURE 5-21 Group Orbitals 1 and 2 for CO_2.

Group orbitals 3 and 4 in Figure 5-22, formed by adding and subtracting the oxygen $2p_z$ orbitals, have the same A_g and B_{1u} symmetries. As in the first two, group orbital 3 can interact with the $2s$ of carbon and group orbital 4 can interact with the carbon $2p_z$.

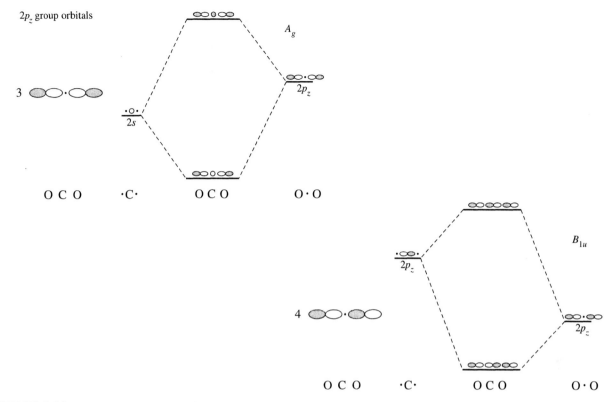

2p_z group orbitals

3

A_g

$2p_z$

$\cdot O \cdot$
$2s$

2p_z

O C O $\cdot$C$\cdot$ O C O   O $\cdot$ O

B_{1u}

$2p_z$

4

$2p_z$

O C O $\cdot$C$\cdot$ O C O   O $\cdot$ O

FIGURE 5-22 Group Orbitals 3 and 4 for CO_2.

The 2s and $2p_z$ orbitals of carbon, therefore, have two possible sets of group orbitals with which they may interact. In other words, all four interactions in Figures 5-21 and 5-22 occur, and all four are symmetry allowed. It is then necessary to estimate which interactions can be expected to be the strongest from the potential energies of the 2s and 2p orbitals of carbon and oxygen given in Figure 5-23.

Interactions are strongest for orbitals having similar energies. Both group orbital 1, from the 2s orbitals of the oxygen, and group orbital 3, from the $2p_z$ orbitals, have the proper symmetry to interact with the 2s orbital of carbon. However, the energy match between group orbital 3 and the 2s orbital of carbon is much better (a difference of 3.6 eV) than the energy match between group orbital 1 and the 2s of carbon (a difference of 12.9 eV); therefore, the primary interaction is between the $2p_z$ orbitals of oxygen and

Orbital	2s	2p
Carbon	−19.4 eV	−10.7 eV
Oxygen	−32.4 eV	−15.9 eV

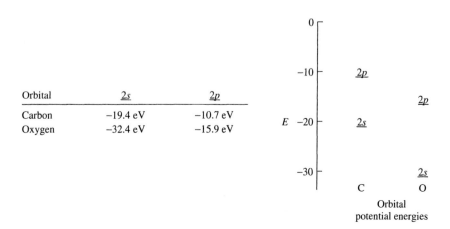

FIGURE 5-23 Orbital Potential
Energies of Carbon and Oxygen.

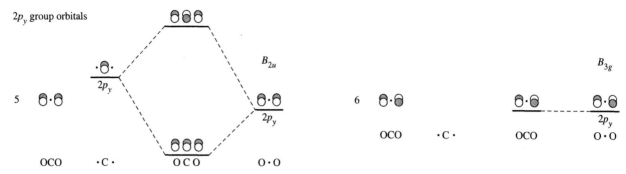

FIGURE 5-24 Group Orbitals 5 and 6 for CO_2.

the $2s$ orbital of carbon. Group orbital 2 also has energy too low for strong interaction with the carbon p_z (a difference of 21.7 eV), so the final molecular orbital diagram (Figure 5-26) shows no interaction with carbon orbitals for group orbitals 1 and 2.

EXERCISE 5-5

Using orbital potential energies, show that group orbital 4 is more likely than group orbital 2 to interact strongly with the $2p_z$ orbital of carbon.

The $2p_y$ orbital of carbon has B_{2u} symmetry and interacts with group orbital 5 (Figure 5-24). The result is the formation of two π molecular orbitals, one bonding and one antibonding. However, there is no orbital on carbon with B_{3g} symmetry to interact with group orbital 6, formed by combining $2p_y$ orbitals of oxygen. Therefore, group orbital 6 is nonbonding.

Interactions of the $2p_x$ orbitals are similar to those of the $2p_y$ orbitals. Group orbital 7, with B_{2u} symmetry, interacts with the $2p_x$ orbital of carbon to form π bonding and antibonding orbitals, whereas group orbital 8 is nonbonding (Figure 5-25).

The overall molecular orbital diagram of CO_2 is shown in Figure 5-26. The 16 valence electrons occupy, from the bottom, two essentially nonbonding σ orbitals, two bonding σ orbitals, two bonding π orbitals, and two nonbonding π orbitals. In other words, two of the bonding electron pairs are in σ orbitals and two are in π orbitals, and there are four bonds in the molecule, as expected. As in the FHF$^-$ case, all the occupied molecular orbitals are 3-center, 2-electron orbitals and all are more stable (have lower energy) than 2-center orbitals.

The molecular orbital picture of other linear triatomic species, such as N_3^-, CS_2, and OCN^-, can be determined similarly. Likewise, the molecular orbitals of longer

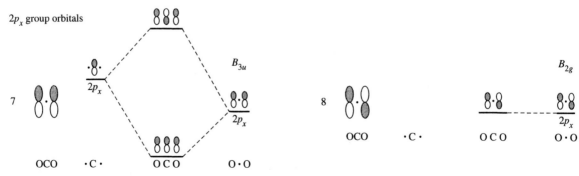

FIGURE 5-25 Group Orbitals 7 and 8 for CO_2.

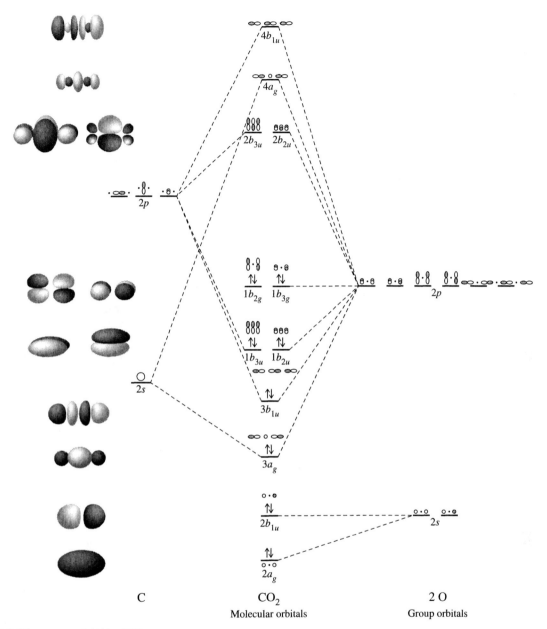

C CO_2 2 O

Molecular orbitals   Group orbitals

FIGURE 5-26 Molecular Orbitals of CO_2.

polyatomic species can be described by a similar method. Examples of bonding in linear π systems will be considered in Chapter 13.

EXERCISE 5-6

Prepare a molecular orbital diagram for the azide ion, N_3^-.

EXERCISE 5-7

Prepare a molecular orbital diagram for the BeH_2 molecule.

5-4-3 H_2O

Molecular orbitals of nonlinear molecules can be determined by the same procedures. Water will be used as an example, and the steps of the previous section will be used.

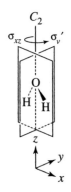

C_2

σ_{xz} σ_v'

O

H   H

z

y

x

FIGURE 5-27 Symmetry of the Water Molecule.

1. Water is a simple triatomic bent molecule with a C_2 axis through the oxygen and two mirror planes that intersect in this axis, as shown in Figure 5-27. The point group is therefore C_{2v}.

2. The C_2 axis is chosen as the z axis and the xz plane as the plane of the molecule.[21] Because the hydrogen $1s$ orbitals have no directionality, it is not necessary to assign axes to the hydrogens.

3. Because the hydrogen atoms determine the symmetry of the molecule, we will use their orbitals as a starting point. The characters for each operation for the $1s$ orbitals of the hydrogen atoms can be obtained easily. The sum of the contributions to the character ($1, 0$, or -1, as described previously) for each symmetry operation is the character for that operation, and the complete list for all operations of the group is the reducible representation for the atomic orbitals. The identity operation leaves both hydrogen orbitals unchanged, with a character of 2. Twofold rotation interchanges the orbitals, so each contributes 0, for a total character of 0. Reflection in the plane of the molecule (σ_v) leaves both hydrogens unchanged, for a character of 2; reflection perpendicular to the plane of the molecule (σ_v') switches the two orbitals, for a character of 0, as in Table 5-2.

TABLE 5-2
Representations for C_{2v} Symmetry Operations for Hydrogen Atoms in Water

C_{2v} Character Table

C_{2v}	E	C_2	$\sigma_v(xz)$	$\sigma_v'(yz)$		
A_1	1	1	1	1	z	x^2, y^2, z^2
A_2	1	1	-1	-1	R_z	xy
B_1	1	-1	1	-1	x, R_y	xz
B_2	1	-1	-1	1	y, R_x	yz

$$\begin{bmatrix} H_a' \\ H_b' \end{bmatrix} = \begin{bmatrix} 1 & 0 \\ 0 & 1 \end{bmatrix} \begin{bmatrix} H_a \\ H_b \end{bmatrix} \text{ for the identity operation}$$

$$\begin{bmatrix} H_a' \\ H_b' \end{bmatrix} = \begin{bmatrix} 0 & 1 \\ 1 & 0 \end{bmatrix} \begin{bmatrix} H_a \\ H_b \end{bmatrix} \text{ for the } C_{2v} \text{ operation}$$

$$\begin{bmatrix} H_a' \\ H_b' \end{bmatrix} = \begin{bmatrix} 1 & 0 \\ 0 & 1 \end{bmatrix} \begin{bmatrix} H_a \\ H_b \end{bmatrix} \text{ for the } \sigma_v \text{ reflection } (xz \text{ plane})$$

$$\begin{bmatrix} H_a' \\ H_b' \end{bmatrix} = \begin{bmatrix} 0 & 1 \\ 1 & 0 \end{bmatrix} \begin{bmatrix} H_a \\ H_b \end{bmatrix} \text{ for the } \sigma_v' \text{ reflection } (yz \text{ plane})$$

The reducible representation $\Gamma = A_1 + B_1$:

C_{2v}	E	C_2	$\sigma_v(xz)$	$\sigma_v'(yz)$	
Γ	2	0	2	0	
A_1	1	1	1	1	z
B_1	1	-1	1	-1	x

[21]Some sources use the yz plane as the plane of the molecule. This convention results in $\Gamma = A_1 + B_2$ and switches the b_1 and b_2 labels of the molecular orbitals.

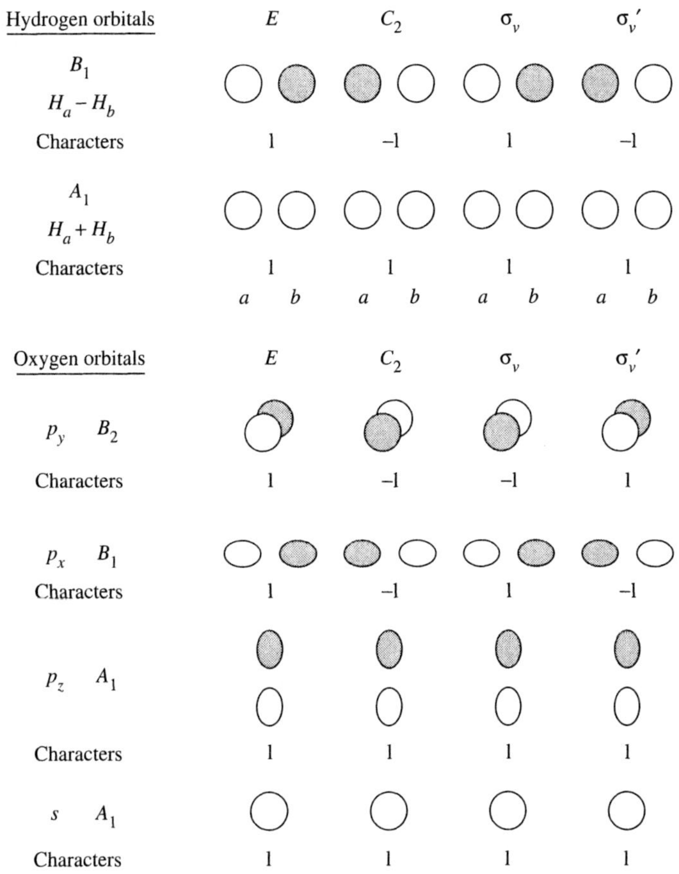

Hydrogen orbitals	E	C_2	σ_v	σ_v'
B_1 $H_a - H_b$				
Characters	1	−1	1	−1
A_1 $H_a + H_b$				
Characters	1	1	1	1
	$a \quad b$	$a \quad b$	$a \quad b$	$a \quad b$

Oxygen orbitals	E	C_2	σ_v	σ_v'
$p_y \quad B_2$				
Characters	1	−1	−1	1
$p_x \quad B_1$				
Characters	1	−1	1	−1
$p_z \quad A_1$				
Characters	1	1	1	1
$s \quad A_1$				
Characters	1	1	1	1

FIGURE 5-28 Symmetry of Atomic and Group Orbitals in the Water Molecule.

4. The representation Γ can be reduced to the irreducible representations $A_1 + B_1$, representing the symmetries of the group orbitals. These group orbitals can now be matched with orbitals of matching symmetries on oxygen. Both $2s$ and $2p_z$ orbitals have A_1 symmetry, and the $2p_x$ orbital has B_1 symmetry. In finding molecular orbitals, the first step is to combine the two hydrogen $1s$ orbitals. The sum of the two, $\frac{1}{\sqrt{2}}[\Psi(H_a) + \Psi(H_b)]$, has symmetry A_1 and the difference, $\frac{1}{\sqrt{2}}[\Psi(H_a) - \Psi(H_b)]$, has symmetry B_1, as can be seen by examining Figure 5-28. These group orbitals, or symmetry-adapted linear combinations, are each then treated as if they were atomic orbitals. In this case, the atomic orbitals are identical and have equal coefficients, so they contribute equally to the group orbitals. The normalizing factor is $\frac{1}{\sqrt{2}}$. In general, the normalizing factor for a group orbital is

$$N = \frac{1}{\sqrt{\Sigma c_i^2}}$$

where c_i = the coefficients on the atomic orbitals. Again, each group orbital is treated as a single orbital in combining with the oxygen orbitals.

5. The same type of analysis can be applied to the oxygen orbitals. This requires only the addition of −1 as a possible character when a p orbital changes sign. Each orbital can be treated independently.

TABLE 5-3
Molecular Orbitals for Water

Symmetry	Molecular Orbitals		Oxygen Atomic Orbitals		Group Orbitals from Hydrogen Atoms	Description
B_1	Ψ_6	$=$	$c_9\,\psi(p_x)$	$+$	$c_{10}\,[\psi(H_a) - \psi(H_b)]$	antibonding (c_{10} is negative)
A_1	Ψ_5	$=$	$c_7\,\psi(s)$	$+$	$c_8\,[\psi(H_a) + \psi(H_b)]$	antibonding (c_8 is negative)
B_2	Ψ_4	$=$	$\psi(p_y)$			nonbonding
A_1	Ψ_3	$=$	$c_5\,\psi(p_z)$	$+$	$c_6\,[\psi(H_a) + \psi(H_b)]$	nearly nonbonding (slightly bonding; c_6 is very small)
B_1	Ψ_2	$=$	$c_3\,\psi(p_x)$	$+$	$c_4\,[\psi(H_a) - \psi(H_b)]$	bonding (c_4 is positive)
A_1	Ψ_1	$=$	$c_1\,\psi(s)$	$+$	$c_2\,[\psi(H_a) + \psi(H_b)]$	bonding (c_2 is positive)

The s orbital is unchanged by all the operations, so it has A_1 symmetry.

The p_x orbital has the B_1 symmetry of the x axis.

The p_y orbital has the B_2 symmetry of the y axis.

The p_z orbital has the A_1 symmetry of the z axis.

The x, y, and z variables and the more complex functions in the character tables assist in assigning representations to the atomic orbitals.

6. The atomic and group orbitals with the same symmetry are combined into molecular orbitals, as listed in Table 5-3 and shown in Figure 5-29. They are numbered Ψ_1 through Ψ_6 in order of their energy, with 1 the lowest and 6 the highest.

The A_1 group orbital combines with the s and p_z orbitals of the oxygen to form three molecular orbitals: one bonding, one nearly nonbonding (slightly bonding), and one antibonding (three atomic or group orbitals forming three molecular orbitals, Ψ_1, Ψ_3, and Ψ_5). The oxygen p_z has only minor contributions from the other orbitals in the weakly bonding Ψ_3 orbital, and the oxygen s and the hydrogen group orbitals combine weakly to form bonding and antibonding Ψ_1 and Ψ_5 orbitals that are changed only slightly from the atomic orbital energies.

The hydrogen B_1 group orbital combines with the oxygen p_x orbital to form two MOs, one bonding and one antibonding (Ψ_2 and Ψ_6). The oxygen p_y (Ψ_4, with B_2 symmetry) does not have the same symmetry as any of the hydrogen $1s$ group orbitals, and is a nonbonding orbital. Overall, there are two bonding orbitals, two nonbonding or nearly nonbonding orbitals, and two antibonding orbitals. The oxygen $2s$ orbital is nearly 20 eV below the hydrogen orbitals in energy, so it has very little interaction with them. The oxygen $2p$ orbitals are a good match for the hydrogen $1s$ energy, allowing formation of the bonding b_1 and a_1 molecular orbitals.

When the eight valence electrons available are added, there are two pairs in bonding orbitals and two pairs in nonbonding orbitals, which are equivalent to the two bonds and two lone pairs of the Lewis electron-dot structure. The lone pairs are in molecular orbitals, one b_2 from the p_y of the oxygen, the other a_1 from a combination of s and p_z of the oxygen and the two hydrogen $1s$ orbitals. The resulting molecular orbital diagram is shown in Figure 5-29.

The molecular orbital picture differs from the common conception of the water molecule as having two equivalent lone electron pairs and two equivalent O—H bonds. In the MO picture, the highest energy electron pair is truly nonbonding, occupying the $2p_y$ orbital perpendicular to the plane of the molecule. The next two pairs are bonding pairs, resulting from overlap of the $2p_z$ and $2p_x$ orbital with the $1s$ orbitals of the hydrogens. The lowest energy pair is a lone pair in the essentially unchanged $2s$ orbital of the oxygen. Here, all four occupied molecular orbitals are different.

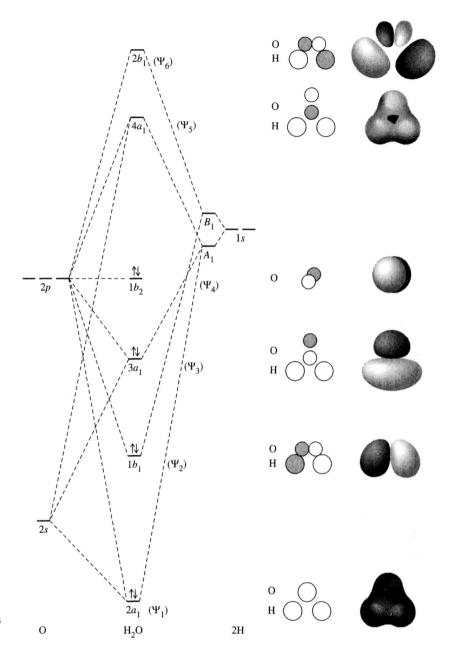

FIGURE 5-29 Molecular Orbitals of H_2O.

O H_2O 2H

5-4-4 NH$_3$

Valence shell electron pair repulsion (VSEPR) arguments describe ammonia as a pyramidal molecule with a lone pair of electrons and C_{3v} symmetry. For the purpose of obtaining a molecular orbital picture of NH_3, it is convenient to view this molecule looking down on the lone pair (down the C_3, or z, axis) and with the the yz plane passing through one of the hydrogens. The reducible representation for the three H atom $1s$ orbitals is given in Table 5-4. It can be reduced by the methods given in Chapter 4 to the A_1 and E irreducible representations, with the orbital combinations in Figure 5-30. Because three hydrogen $1s$ orbitals are to be considered, there must be three group orbitals formed from them, one with A_1 symmetry and two with E symmetry.

TABLE 5-4
Representations for Atomic Orbitals in Ammonia

C_{3v} Character Table

C_{3v}	E	$2 C_3$	$3 \sigma_v$		
A_1	1	1	1	z	$x^2 + y^2, z^2$
A_2	1	1	-1		
E	2	-1	0	$(x, y), (R_x, R_y)$	$(x^2 - y^2, xy)\ (xz, yz)$

The reducible representation $\Gamma = A_1 + E$:

C_{3v}	E	$2 C_3$	$3 \sigma_v$		
Γ	3	0	1		
A_1	1	1	1	z	$x^2 + y^2, z^2$
E	2	-1	0	$(x, y,), (R_x, R_y)$	$(x^2 - y^2, xy)\ (xz, yz)$

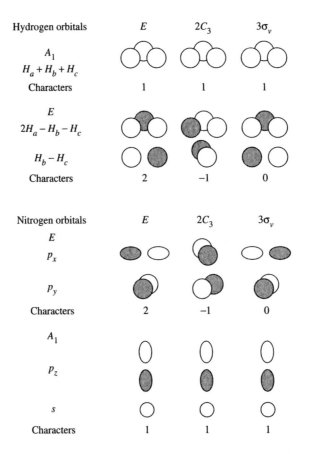

FIGURE 5-30 Group Orbitals of NH_3.

The s and p_z orbitals of nitrogen both have A_1 symmetry, and the pair p_x, p_y has E symmetry, exactly the same as the representations of the hydrogen $1s$ orbitals. Therefore, all orbitals of nitrogen are capable of combining with the hydrogen orbitals. As in water, the orbitals are grouped by symmetry and then combined.

Up to this point, it has been a simple matter to obtain a description of the group orbitals. Each polyatomic example considered (FHF^-, CO_2, H_2O) has had two atoms attached to a central atom and the group orbitals could be obtained by matching atomic orbitals on the terminal atoms in both a bonding and antibonding sense. In NH_3, this is no longer possible. The A_1 symmetry of the sum of the three hydrogen $1s$ orbitals is easily seen, but the two group orbitals of E symmetry are more difficult to see. (The matrix description of C_3 rotation for the x and y axes in Section 4-3-3 may also be helpful.) One condition of the equations describing the molecular orbitals is that the sum of the squares of the coefficients of each of the atomic orbitals in the LCAOs equals 1 for each atomic orbital. A second condition is that the symmetry of the central atom orbitals matches the symmetry of the group orbitals with which they are combined. In this case, the E symmetry of the SALCs must match the E symmetry of the nitrogen p_x, p_y group orbitals that are being combined. This condition requires one node for each of the E group orbitals. With three atomic orbitals, the appropriate combinations are then

$$\frac{1}{\sqrt{6}}[2\Psi(H_a) - \Psi(H_b) - \Psi(H_c)] \quad \text{and} \quad \frac{1}{\sqrt{2}}[\Psi(H_b) - \Psi(H_c)]$$

The coefficients in these group orbitals result in equal contribution by each atomic orbital when each term is squared (as is done in calculating probabilities) and the terms for each orbital summed.

For H_a, the contribution is $\left(\dfrac{2}{\sqrt{6}}\right)^2 = \dfrac{2}{3}$

For H_b and H_c, the contribution is $\left(\dfrac{1}{\sqrt{6}}\right)^2 + \left(\dfrac{1}{\sqrt{2}}\right)^2 = \dfrac{2}{3}$

H_a, H_b, and H_c each also have a contribution of $\frac{1}{3}$ in the A_1 group orbital,

$$\frac{1}{\sqrt{3}}[\Psi(H_a) + \Psi(H_b) + \Psi(H_c)], \quad \left(\frac{1}{\sqrt{3}}\right)^2 = \frac{1}{3}$$

giving a total contribution of 1 by each of the atomic orbitals.

Again, each group orbital is treated as a single orbital, as shown in Figures 5-30 and 5-31, in combining with the nitrogen orbitals. The nitrogen s and p_z orbitals combine with the hydrogen A_1 group orbital to give three a_1 orbitals, one bonding, one nonbonding, and one antibonding. The nonbonding orbital is almost entirely nitrogen p_z, with the nitrogen s orbital combining effectively with the hydrogen group orbital for the bonding and antibonding orbitals.

The nitrogen p_x and p_y orbitals combine with the E group orbitals

$$\frac{1}{\sqrt{6}}[2\Psi(H_a) - \Psi(H_b) - \Psi(H_c)] \quad \text{and} \quad \frac{1}{\sqrt{2}}[\Psi(H_b) - \Psi(H_c)]$$

to form four e orbitals, two bonding and two antibonding (e has a dimension of 2, which requires a pair of degenerate orbitals). When eight electrons are put into the lowest energy levels, three bonds and one nonbonded lone pair are obtained, as suggested by the Lewis electron-dot structure. The $1s$ orbital energies of the hydrogen atoms match well with the energies of the nitrogen $2p$ orbitals, resulting in large differences between the bonding and antibonding orbital energies. The nitrogen $2s$ has an energy low enough that its interaction with the hydrogen orbitals is quite small and the molecular orbital has nearly the same energy as the nitrogen $2s$ orbital.

The HOMO of NH_3 is slightly bonding, because it contains an electron pair in an orbital resulting from interaction of the $2p_z$ orbital of nitrogen with the $1s$ orbitals of the hydrogens (from the zero-node group orbital). This is the lone pair of the electron-dot and VSEPR models. It is also the pair donated by ammonia when it functions as a Lewis base (discussed in Chapter 6).

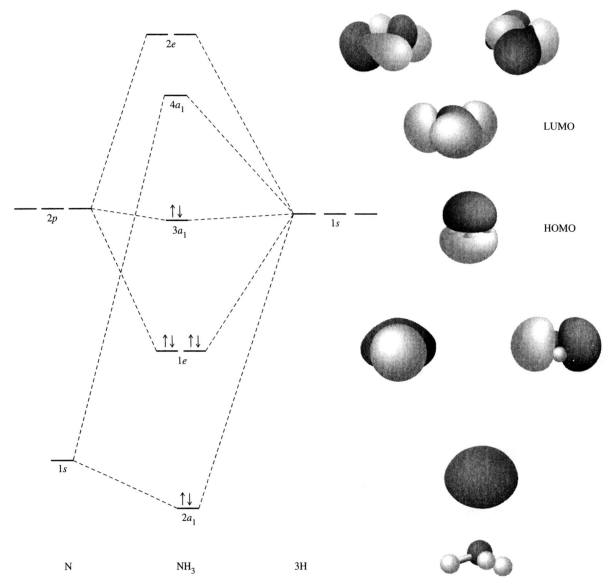

FIGURE 5-31 Molecular Orbitals of NH_3. All are shown with the orientation of the molecule at the bottom.

5-4-5 BF₃

Boron trifluoride is a classic Lewis acid. Therefore, an accurate molecular orbital picture of BF_3 should show, among other things, an orbital capable of acting as an electron pair acceptor. The VSEPR shape is a planar triangle, consistent with experimental observations.

Although both molecules have threefold symmetry, the procedure for describing molecular orbitals of BF_3 differs from NH_3, because the fluorine atoms surrounding the central boron atom have $2p$ as well as $2s$ electrons to be considered. In this case, the p_y axes of the fluorine atoms are chosen so that they are pointing toward the boron atom and the p_x axes are in the plane of the molecule. The group orbitals and their symmetry in the D_{3h} point group are shown in Figure 5-32. The molecular orbitals are shown in

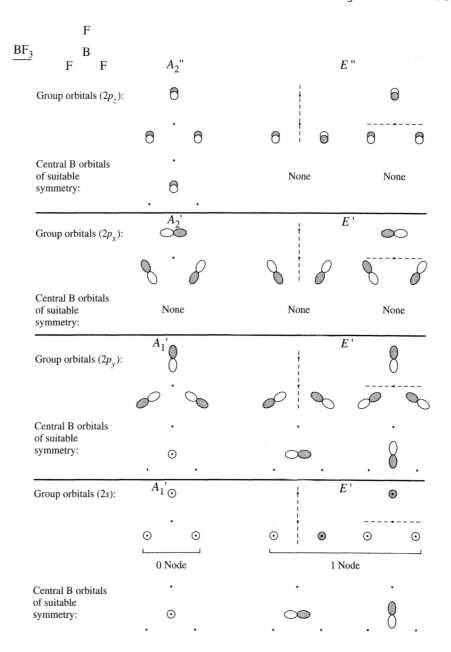

FIGURE 5-32 Group Orbitals for BF$_3$.

Figure 5-33 (omitting sketches of the five nonbonding 2p group orbitals of the fluorine atoms for clarity).

As discussed in Chapter 3, resonance structures may be drawn for BF$_3$ showing this molecule to have some double-bond character in the B—F bonds. The molecular orbital view of BF$_3$ has an electron pair in a bonding π orbital with a_2'' symmetry delocalized over all four atoms (this is the orbital slightly below the five nonbonding electron pairs in energy). Overall, BF$_3$ has three bonding σ orbitals (a_1' and e') and one slightly bonding π orbital (a_2'') occupied by electron pairs, together with eight nonbonding pairs on the fluorine atoms. The greater than 10 eV difference between the B and F p orbital energies means that this π orbital is only slightly bonding.

The LUMO of BF$_3$ is an empty π orbital (a_2''), which has antibonding interactions between the 2p_z orbital on boron and the 2p_z orbitals of the surrounding fluorines.

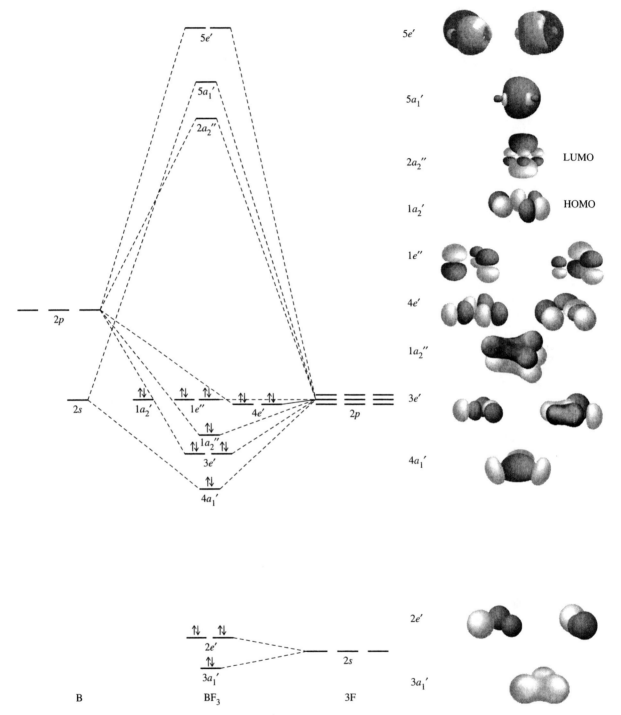

FIGURE 5-33 Molecular Orbitals of BF_3.

This orbital can act as an electron-pair acceptor (for example, from the HOMO of NH_3) in Lewis acid-base interactions.

The molecular orbitals of other trigonal species can be treated by similar procedures. The planar trigonal molecules SO_3, NO_3^-, and CO_3^{2-} are isoelectronic with BF_3, with three σ bonds and one π bond, as expected. Group orbitals can also be used to derive molecular orbital descriptions of more complicated molecules. The simple approach described in these past few pages with minimal use of group theory can lead conveniently to a qualitatively useful description of bonding in simple molecules. More advanced methods based on computer calculations are necessary to deal with more complex molecules and to obtain wave equations for the molecular orbitals. These more advanced methods often use molecular symmetry and group theory.

The qualitative methods described do not allow us to determine the energies of the molecular orbitals, but we can place them in approximate order from their shapes and the expected overlap. The intermediate energy levels in particular are difficult to place in order. Whether an individual orbital is precisely nonbonding, slightly bonding, or slightly antibonding is likely to make little difference in the overall energy of the molecule. Such intermediate orbitals can be described as essentially nonbonding.

Differences in energy between two clearly bonding orbitals are likely to be more significant in the overall energy of a molecule. The π interactions are generally weaker than σ interactions, so a double bond made up of one σ orbital and one π orbital is not twice as strong as a single bond. In addition, single bonds between the same atoms may have widely different energies. For example, the $C-C$ bond is usually described as having an energy near 345 kJ/mol, a value averaged from a large number of different molecules. These individual values may vary tremendously, with some as low as 63 and others as high as 628 kJ/mol.[22] The low value is for hexaphenyl ethane $((C_6H_5)_3C-C(C_6H_5)_3)$ and the high is for diacetylene ($H-C\equiv C-C\equiv C-H$), which are examples of extremes in steric crowding and bonding, respectively, on either side of the $C-C$ bond.

5-4-6 MOLECULAR SHAPES

We used simple electron repulsion arguments to determine the shapes of molecules in Chapter 3, and assumed that we knew the shapes of the molecules described in this chapter. How can we determine the shapes of molecules from a molecular orbital approach? The method is simple in concept, but requires the use of molecular modeling software on a computer to make it a practical exercise.

There are several approaches to the calculation of molecular orbitals. Frequently, the actual calculation is preceded by a simple determination of the shape based on semi-empirical arguments similar to those used in Chapter 3. With the shape determined, the calculations can proceed to determine the energies and compositions of the molecular orbitals. In other cases, an initial estimate of the shape is made and then the two calculations are combined. By calculating the overall energy at different bond distances and angles, the minimum energy is found. One of the principles of quantum mechanics is that any energy calculated will be equal to or greater than the true energy, so we can be confident that the energy calculated is not below the true value.

5-4-7 HYBRID ORBITALS

It is sometimes convenient to label the atomic orbitals that combine to form molecular orbitals as **hybrid orbitals**, or **hybrids**. In this method, the orbitals of the central atom are combined into equivalent hybrids. These hybrid orbitals are then used to form bonds with other atoms whose orbitals overlap properly. This approach is not essential in describing bonding, but was developed as part of the valence bond approach to bonding to describe equivalent bonds in a molecule. Its use is less common today, but it is included here because it has been used so much in the past and still appears in the literature. It has the advantage of emphasizing the overall symmetry of molecules, but is not commonly used in calculating molecular orbitals today.

[22]S. W. Benson, *J. Chem. Educ.*, **1965**, *42*, 502.

Hybrid orbitals are localized in space and are directional, pointing in a specific direction. In general, these hybrids point from a central atom toward surrounding atoms or lone pairs. Therefore, the symmetry properties of a set of hybrid orbitals will be identical to the properties of a set of vectors with origins at the nucleus of the central atom of the molecule and pointing toward the surrounding atoms.

For methane, the vectors point at the corners of a tetrahedron or at alternate corners of a cube. Using the T_d point group, we can use these four vectors as the basis of a reducible representation. As usual, the character for each vector is 1 if it remains unchanged by the symmetry operation, and 0 if it changes position in any other way (reversing direction is not an option for hybrids). The reducible representation for these four vectors is then $\Gamma = A_1 + T_2$.

T_d	E	$8\,C_3$	$3\,C_2$	$6\,S_4$	$6\,\sigma_d$		
Γ	4	1	0	0	2		
A_1	1	1	1	1	1		$x^2 + y^2 + z^2$
T_2	3	0	-1	-1	1	(x, y, z)	(xy, xz, yz)

A_1, the totally symmetric representation, has the same symmetry as the $2s$ orbital of carbon, and T_2 has the same symmetry as the three $2p$ orbitals taken together or the d_{xy}, d_{xz}, and d_{yz} orbitals taken together. Because the d orbitals of carbon are at much higher energy and are therefore a poor match for the energies of the $1s$ orbitals of the hydrogens, the hybridization for methane must be sp^3, combining all four atomic orbitals (one s and three p) into four equivalent hybrid orbitals, one directed toward each hydrogen atom.

Ammonia fits the same pattern. Bonding in NH_3 uses all the nitrogen valence orbitals, so the hybrids are sp^3, including one s orbital and all three p orbitals, with overall tetrahedral symmetry. The predicted HNH angle is $109.5°$, narrowed to the actual $106.6°$ by repulsion from the lone pair, which also occupies an sp^3 orbital.

There are two alternative approaches to hybridization for the water molecule. For example, the electron pairs around the oxygen atom in water can be considered as having nearly tetrahedral symmetry (counting the two lone pairs and the two bonds equally). All four valence orbitals of oxygen are used, and the hybrid orbitals are sp^3. The predicted bond angle is then the tetrahedral angle of $109.5°$ compared with the experimental value of $104.5°$. Repulsion by the lone pairs, as described in the VSEPR section of Chapter 3, is one explanation for this smaller angle.

In the other approach, which is closer to the molecular orbital description of Section 5-4-3, the bent planar shape indicates that the oxygen orbitals used in molecular orbital bonding in water are the $2s$, $2p_x$, and $2p_z$ (in the plane of the molecule). As a result, the hybrids could be described as sp^2, a combination of one s orbital and two p orbitals. Three sp^2 orbitals have trigonal symmetry and a predicted HOH angle of $120°$, considerably larger than the experimental value. Repulsion by the lone pairs on the oxygen (one in an sp^2 orbital, one in the remaining p_y orbital) forces the angle to be smaller.

Similarly, CO_2 uses sp hybrids and SO_3 uses sp^2 hybrids. Only the σ bonding is considered when determining the orbitals used in hybridization; p orbitals not used in the hybrids are available for π interactions. The number of atomic orbitals used in the hybrids is frequently the same as the number of directions counted in the VSEPR method. All these hybrids are summarized in Figure 5-34, along with others that use d orbitals.

Both the simple descriptive approach and the group theory approach to hybridization are used in the following example.

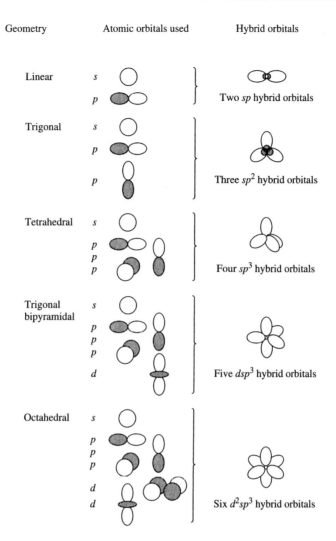

Geometry Atomic orbitals used Hybrid orbitals

FIGURE 5-34 Hybrid Orbitals. Each single hybrid has the general shape ⬭. The figures here show all the resulting hybrids combined, omitting the smaller lobe in the sp^3 and higher orbitals.

EXAMPLE

Determine the types of hybrid orbitals for boron in BF_3.

For a trigonal planar molecule such as BF_3, the orbitals likely to be involved in bonding are the $2s$, $2p_x$, and $2p_y$ orbitals. This can be confirmed by finding the reducible representation in the D_{3h} point group of vectors pointing at the three fluorines and reducing it to the irreducible representations. The procedure for doing this is outlined below.

1. Determine the shape of the molecule by VSEPR techniques and consider each sigma bond to the central atom and each lone pair on the central atom to be a vector pointing out from the center.

2. Find the reducible representation for the vectors, using the appropriate group and character table, and find the irreducible representations that combine to form the reducible representation.

3. The atomic orbitals that fit the irreducible representations are those used in the hybrid orbitals.

Using the symmetry operations of the D_{3h} group, we find that the reducible representation $\Gamma = A_1' + E'$.

D_{3h}	E	$2C_3$	$3C_2$	σ_h	$2S_3$	$3\sigma_v$		
Γ	3	0	1	3	0	1		
A_1'	1	1	1	1	1	1		$x^2 + y^2, z^2$
E'	2	-1	0	2	-1	0	(x, y)	$(x^2 - y^2, xy)$

This means that the atomic orbitals in the hybrids must have the same symmetry properties as A_1' and E'. More specifically, it means that one orbital must have the same symmetry as A_1' (which is one-dimensional) and two orbitals must have the same symmetry, collectively, as E' (which is two-dimensional). This means that we must select one orbital with A_1 symmetry and one *pair* of orbitals that collectively have E' symmetry. Looking at the functions listed for each in the right-hand column of the character table, we see that the s orbital (not listed, but understood to be present for the totally symmetric representation) and the d_{z^2} orbitals match the A_1' symmetry. However, the $3d$ orbitals, the lowest possible d orbitals, are too high in energy for bonding in BF_3 compared with the $2s$ and $2p$. Therefore, the $2s$ orbital is the contributor, with A_1' symmetry.

The functions listed for E' symmetry match the (p_x, p_y) set or the $(d_{x^2-y^2}, d_{xy})$ set. The d orbitals are too high in energy for effective bonding, so the $2p_x$ and $2p_y$ orbitals are used by the central atom. A combination of one p orbital and one d orbital cannot be chosen because orbitals in parentheses must always be taken together.

Overall, the orbitals used in the hybridization are the $2s$, $2p_x$, and $2p_y$ orbitals of boron, comprising the familiar sp^2 hybrids. The difference between this approach and the molecular orbital approach is that these orbitals are combined to form the hybrids before considering their interactions with the fluorine orbitals. Because the overall symmetry is trigonal planar, the resulting hybrids must have that same symmetry, so the three sp^2 orbitals point at the three corners of a planar triangle, and each interacts with a fluorine p orbital to form the three σ bonds. The energy level diagram is similar to the diagram in Figure 5-33, but the three σ orbitals and the three σ^* orbitals each form degenerate sets. The $2p_z$ orbital is not involved in the bonding and serves as an acceptor in acid-base reactions.

EXERCISE 5-8

Determine the types of hybrid orbitals that are consistent with the symmetry of the central atom in

a. PF_5

b. $[PtCl_4]^{2-}$, a square planar ion

The procedure just described for determining hybrids is very similar to that used in finding the molecular orbitals. Hybridization uses vectors pointing toward the outlying atoms and usually deals only with σ bonding. Once the σ hybrids are known, π bonding is easily added. It is also possible to use hybridization techniques for π bonding, but that approach will not be discussed here.[23] Hybridization may be quicker than the molecular orbital approach because the molecular orbital approach uses all the atomic orbitals of the atoms and includes both σ and π bonding directly. Both methods are useful and the choice of method depends on the particular problem and personal preference.

EXERCISE 5-9

Find the reducible representation for all the σ bonds, reduce it to its irreducible representations, and determine the sulfur orbitals used in bonding for $SOCl_2$.

[23]F. A. Cotton, *Chemical Applications of Group Theory*, 3rd ed., John Wiley & Sons, New York, 1990, pp. 227–230.

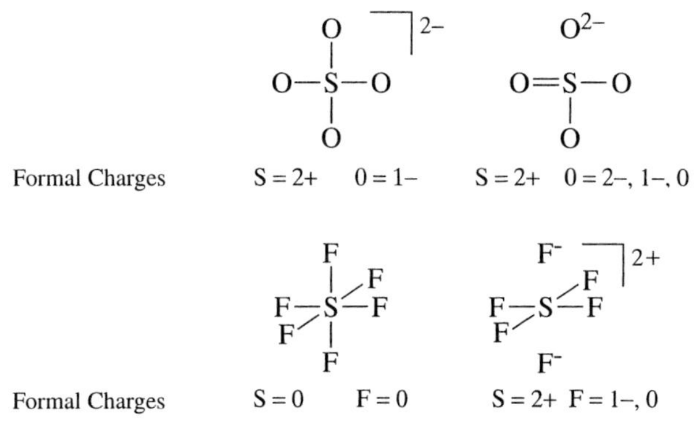

$$
\begin{array}{cc}
\left[\begin{array}{c} O \\ | \\ O-S-O \\ | \\ O \end{array}\right]^{2-} & \begin{array}{c} O^{2-} \\ | \\ O=S-O \\ | \\ O \end{array}
\end{array}
$$

Formal Charges   S = 2+   O = 1–   S = 2+   O = 2–, 1–, 0

FIGURE 5-35 Sulfate and Sulfur Hexafluoride as Described by the Natural Orbital Method.

$$
\begin{array}{cc}
\begin{array}{c} F \\ | \quad F \\ F-S\!\!\diagup\!\!F \\ F\diagup | \\ F \end{array} &
\left[\begin{array}{c} F^- \\ \quad F \\ F-S\!\!\diagup\!\!F \\ F\diagup \\ F^- \end{array}\right]^{2+}
\end{array}
$$

Formal Charges   S = 0   F = 0   S = 2+   F = 1–, 0

5-5 EXPANDED SHELLS AND MOLECULAR ORBITALS

A few molecules described in Chapter 3 required expanded shells in order to have two electrons in each bond (sometimes called hypervalent or hypercoordinate molecules). In addition, formal charge arguments lead to bonding descriptions that involve more than eight electrons around the central atom, even when there are only three or four outer atoms (see Figure 3-6). For example, we have also described SO_4^{2-} as having two double bonds and two single bonds, with 12 electrons around the sulfur. This has been disputed by theoreticians who use the natural bond orbital or the natural resonance theory method. Their results indicate that the bonding in sulfate is more accurately described as a mixture of a simple tetrahedral ion with all single bonds to all the oxygen atoms (66.2%) and structures with one double bond, two single bonds, and one ionic bond (23.1% total, from 12 possible structures), as in Figure 5-35.[24] Some texts have described SO_2 and SO_3 as having two and three double bonds, respectively, which fit the bond distances (143 pm in each) reported for them. However, the octet structures with only one double bond in each molecule fit the calculations of the natural resonance theory method better.

Molecules such as SF_6, which seems to require the use of d orbitals to provide room for 12 electrons around the sulfur atom, are described instead as having four single S—F bonds and two ionic bonds, or as $(SF_4^{2+})(F^-)_2$, also shown in Figure 5-35.[25] This conclusion is based on calculation of the atomic charges and electron densities for the atoms. The low reactivity of SF_6 is attributed to steric crowding by the six fluorine atoms that prevents attack by other molecules or ions, rather than to strong covalent bonds. These results do not mean that we should completely abandon the descriptions presented previously, but that we should be cautious about using oversimplified descriptions. They may be easier to describe and understand, but they are frequently less accurate than the more complete descriptions of molecular orbital theory, and there is still discussion about the best model to use for the calculations. In spite of the remarkable advances in calculation of molecular structures, there is still much to be done.

GENERAL REFERENCES

There are many books describing bonding and molecular orbitals, with levels ranging from those even more descriptive and qualitative than the treatment in this chapter to those designed for the theoretician interested in the latest methods. A classic that starts at the level of this chapter and includes many more details is R. McWeeny's revision of *Coulson's Valence*, 3rd ed., Oxford University Press, Oxford, 1979. A different approach that uses the concept of generator orbitals is that of J. G. Verkade, in *A Pictorial*

[24]L. Suidan, J. K. Badenhoop, E. D. Glendening, and F. Weinhold, *J. Chem. Educ.*, **1995**, *72*, 583.

[25]J. Cioslowski and S. T. Mixon, *Inorg. Chem.*, **1993**, *32*, 3209; E. Magnusson, *J. Am. Chem. Soc.*, **1990**, *112*, 7940.

Approach to Molecular Bonding and Vibrations, 2nd ed., Springer-Verlag, New York, 1997. The group theory approach in this chapter is similar to that of F. A. Cotton, *Chemical Applications of Group Theory*, 3rd ed., John Wiley & Sons, New York, 1990. A more recent book that extends the description is Y. Jean and F. Volatron, *An Introduction to Molecular Orbitals*, translated and edited by J. K. Burdett, Oxford University Press, Oxford, 1993. J. K. Burdett, *Molecular Shapes*, John Wiley & Sons, New York, 1980, and B. M. Gimarc, *Molecular Structure and Bonding*, Academic Press, New York, 1979, are both good introductions to the qualitative molecular orbital description of bonding.

PROBLEMS

5-1   Expand the list of orbitals considered in Figures 5-2 and 5-3 by using all three *p* orbitals of atom A and all five *d* orbitals of atom B. Which of these have the necessary match of symmetry for bonding and antibonding orbitals? These combinations are rarely seen in simple molecules, but can be important in transition metal complexes.

5-2   Compare the bonding in O_2^{2-}, O_2^-, and O_2. Include Lewis structures, molecular orbital structures, bond lengths, and bond strengths in your discussion.

5-3   Although the peroxide ion, O_2^{2-}, and the acetylide ion, C_2^{2-}, have long been known, the diazenide ion (N_2^{2-}) has only recently been prepared. By comparison with the other diatomic species, predict the bond order, bond distance, and number of unpaired electrons for N_2^{2-}. (Reference: G. Auffermann, Y. Prots, and R. Kniep, *Angew. Chem., Int. Ed.*, **2001**, *40*, 547)

5-4   **a.** Prepare a molecular orbital energy level diagram for NO, showing clearly how the atomic orbitals interact to form MOs.

       b. How does your diagram illustrate the difference in electronegativity between N and O?

       c. Predict the bond order and the number of unpaired electrons.

       d. NO^+ and NO^- are also known. Compare the bond orders of these ions with the bond order of NO. Which of the three would you predict to have the shortest bond? Why?

5-5   **a.** Prepare a molecular orbital energy level diagram for the cyanide ion. Use sketches to show clearly how the atomic orbitals interact to form MOs.

       b. What is the bond order, and how many unpaired electrons does cyanide have?

       c. Which molecular orbital of CN^- would you predict to interact most strongly with a hydrogen $1s$ orbital to form an H—C bond in the reaction $CN^- + H^+ \longrightarrow HCN$? Explain.

5-6   The hypofluorite ion, OF^-, can be observed only with difficulty.

       a. Prepare a molecular orbital energy level diagram for this ion.

       b. What is the bond order and how many unpaired electrons are in this ion?

       c. What is the most likely position for adding H^+ to the OF^- ion? Explain your choice.

5-7   Although KrF^+ and XeF^+ have been studied, $KrBr^+$ has not yet been prepared. For $KrBr^+$:

       a. Propose a molecular orbital diagram, showing the interactions of the valence shell *s* and *p* orbitals to form molecular orbitals.

       b. Toward which atom would the HOMO be polarized? Why?

       c. Predict the bond order.

       d. Which is more electronegative, Kr or Br? Explain your reasoning.

5-8   Prepare a molecular orbital energy level diagram for SH^-, including sketches of the orbital shapes and the number of electrons in each of the orbitals. If a program for calculating molecular orbitals is available, use it to confirm your predictions or to explain why they differ.

5-9   Methylene, CH_2, plays an important role in many reactions. One possible structure of methylene is linear.

       a. Construct a molecular orbital energy level diagram for this species. Include sketches of the group orbitals, and indicate how they interact with the appropriate orbitals of carbon.

       b. Would you expect linear methylene to be diamagnetic or paramagnetic?

5-10   In the gas phase, BeF_2 forms linear monomeric molecules. Prepare a molecular orbital energy level diagram for BeF_2, showing clearly which atomic orbitals are involved in bonding and which are nonbonding.

5-11   For the compound XeF_2:

   a. Sketch the valence shell group orbitals for the fluorine atoms (with the z axes collinear with the molecular axis).

   b. For each of the group orbitals, determine which outermost s, p, and d orbitals of xenon are of suitable symmetry for interaction and bonding.

5-12   Prepare a molecular orbital energy level diagram for the ozone molecule, O_3, for each of the following conditions:

   a. Without mixing of the s and p orbitals.

   b. Indicate the changes in molecular orbital energies that you would predict on mixing of s and p orbitals.

5-13   The ion H_3^+ has been observed, but its structure has been the subject of some controversy. Prepare a molecular orbital energy level diagram for H_3^+, assuming a cyclic structure. (The same problem for a linear structure is given in Exercise 5-4.)

5-14   Describe the bonding in SO_3 by using group theory to find the molecular orbitals. Include both the σ and π orbitals, and try to put the resulting orbitals in approximate order of energy. (The actual results are more complex because of mixing of orbitals, but a simple description can be found by the methods given in this chapter.)

5-15   Use molecular orbital arguments to explain the structures of SCN^-, OCN^-, and CNO^- and compare the results with the electron-dot pictures of Chapter 3.

5-16   Thiocyanate and cyanate ions both bond to H^+ through the nitrogen atoms (HNCS and HNCO), whereas SCN^- forms bonds with metal ions through either nitrogen or sulfur, depending on the rest of the molecule. What does this suggest about the relative importance of S and N orbitals in the MOs of SCN^-? (Hint: See the discussion of CO_2 bonding.)

5-17   The thiocyanate ion, SCN^-, can form bonds to metals through either S or N (Problem 5-16). What is the likelihood of cyanide, CN^-, forming bonds to metals through N as well as C?

5-18   The isomeric ions NSO^- (thiazate) and SNO^- (thionitrite) ions have been reported by S. P. So, *Inorg. Chem.*, **1989**, *28*, 2888.

   a. On the basis of the resonance structures of these ions, predict which would be more stable.

   b. Sketch the approximate shapes of the π and π^* orbitals of these ions.

   c. Predict which ion would have the shorter N—S bond and which would have the higher energy N—S stretching vibration? (Stronger bonds have higher energy vibrations.)

5-19 SF_4 has C_{2v} symmetry. Predict the possible hybridization schemes for the sulfur atom in SF_4.

5-20   Consider a square pyramidal AB_5 molecule. Using the C_{4v} character table, determine the possible hybridization schemes for central atom A. Which of these schemes would you expect to be most likely?

5-21   In coordination chemistry, many square planar species are known (for example, $PtCl_4^{2-}$). For a square planar molecule, use the appropriate character table to determine the types of hybridization possible for a metal surrounded in a square planar fashion by four ligands (consider hybrids used in σ bonding only).

5-22   For the molecule PCl_5:

   a. Using the character table for the point group of PCl_5, determine the possible type(s) of hybrid orbitals that can be used by P in forming σ bonds to the five Cl atoms.

   b. What type(s) of hybrids can be used in bonding to the axial chlorine atoms? To the equatorial chlorine atoms?

   c. Considering your answer to Part b, explain the experimental observation that the axial P—Cl bonds (219 pm) are longer than the equatorial bonds (204 pm).

5-23   Describe the bonding in the sulfite ion, SO_3^{2-}, in terms of the electron-dot pictures of Chapter 3, including reduction of the formal charges as much as possible, then in terms of molecular orbitals, and, finally, using the combined covalent-ionic mixture described by L. Suidan, J. K. Badenhoop, E. D. Glendening, and F. Weinhold, *J. Chem. Educ.*, **1995**, *72*, 583.

5-24   Diborane, B_2H_6, has the structure shown. Using molecular orbitals (and showing appropriate orbitals on B and H from which the MOs are formed), explain how hydrogen can form "bridges" between two B atoms. (This type of bonding is discussed in Chapter 8.)

5-25   Although the Cl_2^+ ion has not been isolated, it has been detected in the gas phase by UV spectroscopy. An attempt to prepare this ion by reaction of Cl_2 with IrF_6 yielded not Cl_2^+, but the rectangular ion Cl_4^+. (Reference: S. Seidel and K. Seppelt, *Angew. Chem., Int. Ed.*, **2000**, *39*, 3923.)

   a. Compare the bond distance and bond energy of Cl_2^+ with Cl_2.

   b. Account for the bonding in Cl_4^+. This ion contains two short Cl—Cl bonds and two much longer ones. Would you expect the shorter Cl—Cl distances in Cl_4^+ to be longer or shorter than the Cl—Cl distance in Cl_2^+? Explain.

5-26 BF_3 is often described as a molecule in which boron is electron deficient, with an electron count of six. However, resonance structures can be drawn in which boron has an octet, with delocalized π electrons.

   a. Draw these structures.

   b. Find the molecular orbital in Figure 5-33 that shows this delocalization and explain your choice.

   c. BF_3 is *the* classic Lewis acid, accepting a pair of electrons from molecules with lone pairs. Find the orbital in Figure 5-33 that is this acceptor and explain your choice, including why it looks like a good electron acceptor.

   d. What is the relationship between the orbitals identified in Parts b and c?

The following problems require the use of molecular modeling software.

5-27   **a.** Identify the point group of the $1a_2''$, $2a_2''$, $1a_2'$, and $1e''$ molecular orbitals in Figure 5-33.

   b. Use molecular modeling software to calculate and view the molecular orbitals BF_3.

   c. Print out the contributions of the atomic orbitals to the atomic orbitals of the $3a_1'$, $4a_1'$, $1a_2''$, $1a_2'$, and $2a_2''$ molecular orbitals, confirming (if you can) the atomic orbital combinations shown in Figure 5-33.

5-28   The ions and molecules NO^+, CN^-, CO, and N_2 form an isoelectronic series. The changing nuclear charges will also change the molecular energy levels of the orbitals formed from the $2p$ atomic orbitals (1π, 3σ, and $1\pi^*$). Use molecular modeling software to

   a. Calculate and display the shapes of these three molecular orbitals for each species (CO and N_2 are included in this chapter).

   b. Compare the shapes of each of the orbitals for each of the species (for example, the shapes of the 1π orbitals for each). What trends do you observe?

   c. Compare the energies of each of the orbitals. For which do you see evidence of mixing?

5-29   Calculate and display the orbitals for the linear molecule BeH_2. Describe how they illustrate the interaction of the outer group orbitals with the orbitals on the central atom.

5-30   Calculate and display the orbitals for the linear molecule BeF_2. Compare the orbitals and their interactions with those of BeH_2 from Problem 5-29. In particular, indicate the outer group orbitals that do not interact with orbitals on the central atom.

5-31   The azide ion, N_3^-, is another linear triatomic species. Calculate and display the orbitals for this ion and compare the three highest energy occupied orbitals with those of BeF_2. How do the outer atom group orbitals differ in their interactions with the central atom orbitals? How do the orbitals compare with the CO_2 orbitals?

5-32   **a.** Calculate and display the molecular orbitals for linear and cyclic H_3^+.

   b. Which species is more likely to exist (i.e., which is more stable)?